ENVIRONMENTAL SCIENCE, ENGINEERING AND TECHNOLOGY

PRECIPITATION

PREDICTION, FORMATION AND ENVIRONMENTAL IMPACT

Environmental Science, Engineering and Technology

Additional books in this series can be found on Nova's website
under the Series tab.

Additional e-books in this series can be found on Nova's website
under the e-book tab.

ENVIRONMENTAL SCIENCE, ENGINEERING AND TECHNOLOGY

PRECIPITATION

PREDICTION, FORMATION AND ENVIRONMENTAL IMPACT

HENRY DOHRING
AND
JEREMY DIXON
EDITORS

Nova Science Publishers, Inc.
New York

For permission to use material from this book please contact us:
Telephone 631-231-7269; Fax 631-231-8175
Web Site: http://www.novapublishers.com

Additional color graphics may be available in the e-book version of this book.

Library of Congress Cataloging-in-Publication Data

Precipitation : prediction, formation, and environmental impact / [edited by] Henry Dohring and Jeremy Dixon.
p. cm.
Includes bibliographical references and index.
ISBN 978-1-62100-447-9 (hardcover)
1. Precipitation forecasting. 2. Precipitation (Meteorology) 3. Precipitation (Chemistry) I. Dohring, Henry, 1972- II. Dixon, Jeremy.
QC925.P665 2011
551.57'7--dc23
2011035496

Published by Nova Science Publishers, Inc. † New York

CONTENTS

PREFACE

In this book, the authors present current research in the study of the prediction, formation and environmental impact of precipitation. Topics discussed in this compilation include the precipitation of anions in aqueous systems; inter-decadal trend turnings of climate modes and dynamic/thermodynamic structures associated with a series of winter precipitation events over China; stochastic modeling and simulation of daily precipitation for weather generation; anion intercalation during Mg-Al LDH precipitation and main water-soluble ions in precipitation of the central Mediterranean region.

Chapter 1 - Anions such as arsenate, borate, fluoride, phosphate, and silicate are of environmental significance. Precipitation-dissolution is an important reaction that governs their fate, transport, and concentration in the environment. Precipitation is also a physico-chemical process in water and wastewater treatment. This paper reviews precipitation of anions, including its chemistry, equilibrium prediction, and the environmental implications as applied in natural systems and wastewater treatment. The application of precipitation process to remove anions from wastewater is assessed. Critical parameters of anion precipitation, such as pH, degree of saturation, metal salts, and interferences of other ions are examined. Thermodynamic modeling can provide the theoretical analysis of main parameters and evaluate the effect of solution condition in precipitation process. The review compares several models, such as GEOCHEM, WATEQ4F, MINTEQ, VIS. MINTEQ, and PHREEQC that are commonly used to predict the speciation and SI (saturation index) in aqueous solution. It is aimed to provide a summary of recent developments in precipitation of anions in aqueous systems.

Chapter 2 - Inter-decadal climate variability is an important reference for inter-annual climate change and is also a disturbance in long term climate change. Inter-decadal climate variability has become a hot topic for climate change studies since the 1980s. In this context, a study of the change in surface air temperature (SAT) and precipitation on an inter-decadal time scale will elucidate the evolution and abrupt changes in the climate. In this chapter, the authors identified the spatial/temporal characteristics of the inter-decadal trend turning (ITT) of winter/summer climate modes over Chinese mainland. The research results show that the time of the "warm-wet" mode turning for winter climate is earlier than that for summer climate. Winter climate exhibits a distinctive "warm-dry" trend in North China after the late 1970s, and a slight "warm-dry" trend in Northeast China, South China, and Southwest China after the late 1980s. But a "warm-wet" trend appears in the rest four sub-areas after the early 1980s. Recently, the summer climate trends towards a "warm-dry" or "warm-wet" mode in most regions of China, except for South China, its summer climate starts a "cold-wet" trend in 1984. Furthermore, an analysis of ITTs of SAT and precipitation in global terrene indicates that warming trends have become highly significant across most of the world in the late 20th Century, and most terrestrial grids also have recorded an ITT of precipitation in either the 1970s or 1980s. Basically, global terrestrial regions have experienced either an increasingly "warm-dry" or an increasingly "warm-wet" climate. The northern American continent has become increasingly warm and dry after the last ITT. Meanwhile, the African continent has become increasingly warm and wet, with both Europe and most of Asia having the same trend. Southern South America and the west of Australia have experienced opposite trends in climate, becoming increasingly cold and wet.

On the other hand, the extreme event under global warming is also a focus in climate research. In winter of 2008, China experienced once-in-50-years (or once-in-100-years for some regions) snow and ice storms. These storms brought huge socio-economical impacts on the Chinese people and government. In this chapter, these cases were revisited and comprehensive analyses of the storms' dynamic/thermodynamic structure were conducted. These snowstorms were also compared with U.S. east coast snowstorms. The results from this study will provide insight on how to improve forecasts for these kinds of snowstorms. The analyses demonstrated that the storms exhibited classic patterns of large-scale circulation common to these types of snowstorms. However, several physical processes were found to be unique, and thought to have played crucial roles in intensifying and prolonging China's great snowstorms of 2008. These include a subtropical high over the west Pacific, an upper-level jet stream, and temperature and moisture inversions. The combined effects of these dynamic/thermodynamic structures are responsible for the development of the storms into a major disastrous event in Chinese history.

Chapter 3 - Weather generation is critical when the historical data is not comprehensive for the required analysis. Therefore, the purpose of weather generation is to generate synthetic sequences of weather variables (e.g., precipitation) that are statistically indistinguishable from the observed characteristics of the historical records. These generated sequences provide a set of alternative realizations for use in the design, operation, management, and policy decisions related to water resources and hydrological systems. The significance of weather generation, for example, can easily be appreciated by examining different research areas, such as modeling of weather, analysis of climate-sensitive systems, and impact assessment of scenarios on water balance in the wake of impending climate change.

Weather generation normally implies simulation of precipitation data, which serves as an important input to hydrological systems and is of importance in local rainfall-runoff studies. To be able to capture the statistical characteristics of the precipitation data into the model, (1) the unique characteristics of the data set, (2) the temporal dependency in a precipitation series, and (3) the spatial dependency of precipitation amount among stations have to be modeled properly.

Due to the unique characteristics of the precipitation data set, modelling of daily precipitation data commonly faces with several challenges, such as: (i) the difficulty of accommodating the occurrence of sustained dry and wet periods; (ii) the prevalence of missing and trace records causing unspecified uncertainty in analyses; (iii) the familiar probability distributions incapability of modelling the precipitation time series, as such series tend to be discontinuous comprised zeros (occurring intermittently) and non-negative numbers.

In respect to the modeling of temporal dependency, precipitation sequences often have been modelled using a short-memory model, but results have not always been satisfactory. Some researchers have advocated the inclusion of higher order memory components for improving the simulation results in terms of long-term dependencies such an approach may result over parameterized the model.

Likewise, the modelling on a spatial basis faces problems because of the occurrence of non-Normal probability distributions of precipitation data. If precipitation data were normally distributed, it would be relatively easy to develop a spatial linkage among the adjoining precipitation stations in the form of multivariate Normal distribution. However, precipitation data have a lower bound of zero and are skewed distributed that makes it difficult to establish a multivariate relationship among the precipitation data of different stations.

Precipitation series have been conventionally regarded as a stationary process. Pragmatically, one cannot ignore the influence of global warming on precipitation, particularly at the present juncture when rising global temperatures are causing alarms and are impacting atmospheric processes. Thus, the influence of climate change to the modeling of precipitation data is drawing more attention for the last few decades. In view of the above challenges, this chapter aims to present the concepts and methods that can preserve the statistical characteristics of precipitation for the weather generation.

Chapter 4 - Magnesium–aluminum layered double hydroxides (Mg–Al LDHs) possess anion-exchange properties that can potentially be utilized for water preservation and purification. Therefore, a simple and convenient method for preparing Mg–Al LDHs is required. In this chapter, the authors developed new preparation methods of Mg–Al LDH by using seawater and $Ca(OH)_2$ as magnesium and alkali resources. Furthermore, dolomite and polyaluminum chloride were examined as potential magnesium, aluminum, and alkali resources for the production of Mg–Al LDHs. In addition, the authors analyzed the formation of Mg–Al LDH, as below. Mg–Al solution, together with alkali (used for pH control), was added dropwise to deionized water. In the local environment, where the droplets of alkali were added, the pH was very high. This high pH resulted in the formation of $Al(OH)_4^-$ from $Al(OH)_3$; this $Al(OH)_3$ formed in solution with the addition of Mg–Al solution. The $Al(OH)_4^-$ quickly reacted with Mg^{2+} and other anions present in the solution, thereby leading to the formation of Mg−Al LDH. Moreover, the authors examined the characteristics of anion intercalation during Mg−Al LDH precipitation and investigated the Mg–Al LDH particle properties for the various preparation methods.

Chapter 5 - The chemical content of atmospheric deposition is the signature of several interacting physical and chemical mechanisms: i.e., emission and source amplitude; transport processes and dynamics of the atmosphere; atmospheric chemical reactions; and removal processes, among others. Studying deposition processes and amounts thus allow for tracing the spatio-temporal evolution of atmospheric chemistry. It is also an important way for distinguishing between natural and anthropogenic influences. Sulfate, nitrate and ammonium ions are formed in the atmosphere in gas-particle conversion processes. Sulfates represent one of the main components of atmospheric aerosols. In southern Europe, rainwater has similar concentrations of acidic anions to those of other parts of Europe, but the average precipitation is not acid due to neutralizing role of basic cations. The Mediterranean Sea is enclosed basin, which is susceptible to dust transport from desert sources in North Africa (Sahara) and the Middle East (the Arabian Peninsula and Syria), lying on its periphery. The climatic conditions are favorable for the transport of dust from North Africa into the eastern Mediterranean atmosphere during the spring, while other conditions favor transport into the western Mediterranean atmosphere during the summertime. An investigation of wet deposition of water soluble Na^+, K^+, Mg^{2+}, Ca^{2+}, Cl^-, SO_4^{2-}, NO_3^- and NH_4^+ were done in south-eastern Adriatic coast in the period from January 1995 until December 2000. The highest precipitation and the highest frequencies of air masses coming from the western Mediterranean cause the highest Wet Deposition (WD) of the main water soluble ions. The most abundant WD of SO_4^{2-} in investigated region was caused by air masses coming from the western Mediterranean and western and central Europe. The sea salt components (Na^+, Cl^-, Mg^{2+}) are significantly correlated. The correlations between Ca^{2+} and Mg^{2+}, Ca^{2+} and SO_4^{2-} and Ca^{2+} and K^+ indicate that the main terrigeneous ions were associated with formation of eolian particles which include gypsum ($CaSO_4$), calcite ($CaCO_3$) and dolomite ($CaMg(CO_3)_2$). Although the Volume Weighted Mean (VWM) of Na^+, K^+, Mg^{2+}, Ca^{2+}, Cl^-, SO_4^{2-}, NO_3^- and NH_4^+ that are coming in air masses from the continental parts of Europe are significant, the total Wet Deposition (WD) is dominated by precipitation coming over Mediterranean due to abundances of precipitation from this segment.

Chapter 6 - Precipitation amount is one of the most demanded meteorological parameters for climate change impact studies. It represents an important input for studies across various economic sectors such as agriculture, hydrology, energy and tourism. Furthermore, precipitation acts as a trigger for extreme events such as landslides and floods. However, the correct simulation of precipitation intensity and the sequence of wet and dry conditions at the daily and sub-daily scale is very challenging. Homogenous Markov chains represent one of the first attempts to capture and reproduce daily precipitation characteristics and are still common practice in statistical modeling. Besides, climate models nowadays are the state-of-the-art tools for dynamically simulating historic as well as future climate conditions in a physically consistent way. Climate models are mathematical representations of the climate system with its subsystems, their processes and interactions.

ACKNOWLEDGMENTS

This work was jointly supported by the Key Project of Nature Science Fund of China (No. 41130960), the International Sci.-Tech. Cooperative Project (2009DFB20540) and the social commonweal profession research program (GYHY201006009) funded by the Ministry of Science and Technology of the People's Republic of China, the Key Project of Basic Scientific Research and Operation fund of the Chinese Academy of Meteorological Sciences (2011Z001).

In: Precipitation: Prediction, Formation, Environmental Impact ISBN: 978-1-62100-447-9
Editors: Henry Dohring and Jeremy Dixon

Chapter 1

PRECIPITATION OF ANIONS: CHEMISTRY, PREDICTION, AND ENVIRONMENTAL APPLICATIONS

Warmadewanthi[1], H. M. Citraningrum[2] and J. C. Liu[2*]
[1]Department of Environmental Engineering,
Institut Teknologi Sepuluh Nopember (ITS), Surabaya, Indonesia
[2]Department of Chemical Engineering,
National Taiwan University of Science and Technology,
Taipei, Taiwan

ABSTRACT

Anions such as arsenate, borate, fluoride, phosphate, and silicate are of environmental significance. Precipitation-dissolution is an important reaction that governs their fate, transport, and concentration in the environment. Precipitation is also a physico-chemical process in water and wastewater treatment. This paper reviews precipitation of anions, including its chemistry, equilibrium prediction, and the environmental implications as applied in natural systems and wastewater treatment. The application of precipitation process to remove anions from wastewater is assessed. Critical parameters of anion precipitation, such as pH, degree of saturation, metal salts, and interferences of other ions are examined. Thermodynamic modeling can provide the theoretical analysis of main parameters and evaluate the effect of solution condition in precipitation process. The review compares several models, such as GEOCHEM, WATEQ4F, MINTEQ, VIS. MINTEQ, and PHREEQC that are commonly used to predict the speciation and SI (saturation index) in aqueous solution. It is aimed to provide a summary of recent developments in precipitation of anions in aqueous systems.

Keywords: Anions, arsenate, borate, fluoride, model, molybdate, phosphate, precipitation, selenium, silica, wastewater, water

---

* Corresponding author at: liu1958@mail.ntust.edu.tw.

1. Anions in Industrial Wastewater

Anions are ubiquitous in aqueous solutions. Some are essential micronutrients, such as phosphate, fluoride, and borate, while some are highly toxic, for example, arsenic. Some anions exist in more than one oxyanion forms, each containing different number of oxygen atoms. The speciation of anions is determined by pH through protonation-deprotonation. pH also significantly affects precipitation-dissolution, which is an important reaction that governs the fate, transport, and concentration of anions in the environment. Even though anions in water and wastewater are mostly treated by conventional techniques, such as adsorption, ion exchange, and etc, it is noted that these methods are most suitable for dilute systems. It has been indicated that high concentration of anions could be found in some industrial wastewater as well as contaminated groundwater. For example, manufacturing process of semiconductor, thin-film transistor liquid crystal display (TFT-LCD), and light emitting diodes (LEDs) produces wastewater that contains a significant amount of fluoride, phosphate, borate, and arsenic. Prior to discharging wastewater to water body, treatment for removal of anions is required. Precipitation process has been demonstrated effective and efficient to remove concentrated anions from wastewater.

Chemical precipitation process is a well established treatment method for water and wastewater treatment. This process can be used for water softening, heavy metal removal, phosphate removal, and other wastewater. The advantages of this method include low cost and low maintenance [USEPA, 2000]. However, large amount of sludge produced must be properly handled. Nowadays, more and more studies have focused on recovery and reuse of sludge as resources. Because of variations in property of aqueous solution, the calculation of optimum dosage is practically impossible and jar-test method is needed to determine the dosage of precipitation reagent. Recently, many equilibrium models in computer software are available for us to better assess chemical speciation, precipitation, and complexation in water and wastewater. Recent developments in using precipitation process for anion removal, applications of computer models in related equilibrium prediction, and the environmental implications are reviewed in this work. It is aimed to provide with a comprehensive introduction for this topic.

1.1. Arsenic

Arsenic exists in nature mostly as arsenic-containing rocks and sediments with typical concentration of 0.5 to 2.5 mg/kg [Mohan and Pittman Jr., 2007]. It has four oxidation forms: As (V), As (III), As (0), and As (-III) [Choong et al., 2007]. When present in oxidizing environment, arsenate (As (V)) is more stable, while arsenite (As (III)) is predominant in reducing environment.

Arsenic in groundwater is mostly contributed by natural mobilization, such as weathering, volcanic eruptions, and geochemical reactions [Mohan and Pittman Jr., 2007]. Apart from natural existence, arsenic also enters the water system from industrial effluents, fossil fuels combustions, and extensive use of pesticides and herbicides. Smelting for metals, such as iron, copper, and zinc contributes to the discharge of arsenic into the water body. A mining site in Mexico showed arsenic concentration of 5.07 mg/L [Song et al., 2006], while as high as 1,628 mg/L of arsenic was found in copper smelting wastewater [Basha et al., 2008]. Wastewater containing 43 mg/L arsenic was produced from hydrometallurgical recovery of germanium from fly ash [Kalderis et al., 2008]. It is also noted that wastewater contained 450-480 mg/L arsenic from sulfuric acid production [Wang et al., 2011]. In high-tech industries such as light emitting diodes (LEDs) production, arsenic in wastewater varied between 479 – 1050 mg/L, and was originated from two sources: chemical vapor deposition of AsH_3 and solid waste from chemical mechanical polishing (CMP) of the final product [Huang et al., 2007].

Arsenic is widely known of its toxicity even at low concentration and its mobility over a wide range of pH. World Health Organization (WHO) sets a standard of 0.01 mg/L (10 ppb) arsenic in drinking water, however, in some countries the permissible limit are often given at 50 ppb [Wyatt et al., 1998; Mandal and Suzuki, 2002; Alam et al., 2003; Kunli et al., 2004]. As for effluent standard of arsenic, each country has its own rule. China sets the limit at 5 mg/L [Wu et al., 2011], while Chile and Taiwan have the same standard of 0.5 mg/L [Chen and Chung, 2006; Nunez et al., 2011].

1.2. Boron

Boron is a common element in the environment, and it exists mainly as borax ($Na_2B_4O_7.10\ H_2O$), datolite ($CaBSiO_4(OH)$), colemanite ($CaB_3O_4(OH)_3.H_2O$), and ulexite ($NaCaB_5O_6(OH)_6.5H_2O$) [Parks and Edwards, 2005]. Boron in the aquatic environment is predominantly in the form of boric acid, $B(OH)_3$, and borate, $B(OH)_4^-$, depending on the pH of aqueous solution and it is widely distributed in the environment naturally or from anthropogenic sources [Xu and Jiang, 2007].

Boron is found among various types of wastewater with different concentration. Optoelectronic wastewater contained high concentrations of boron (> 760 mg/L) and iodine [Mohapatra et al., 2008; Kentjono et al., 2010]. A geothermal power plant discharged wastewater containing up to 24 mg/l boron [Demirel and Yildirim, 2002; Kabay et al., 2004]. Boron reached 119 mg/L in geothermal water in Tibet [Guo et al., 2008]. Ceramic wastewater was found to contain 15 mg/L of boron [Chong et al., 2009]. It ranged from 63.5 to 76.5 mg/L in landfill leachate and can be effectively treated by electrodialytic and nanofiltration method [Turek et al., 2005; Turek et al., 2007]. Groundwater was found to be high in boron and arsenic concentrations because of mining activity in Western Turkey [Gemici et al., 2008]. High concentration of boron has an adverse effect on living organism and wastewater containing boron is a global concern [Parks and Edwards, 2005]. WHO sets the drinking water standard of 0.5 mg/L for boron [WHO, 2003]. The effluent standards for boron in Japan are 10 mg/L and 230 mg/L in non-coastal and coastal area, respectively [Japan MOE, 2007]. In Taiwan, the effluent standard for boron is 1 mg/L [Taiwan EPA, 2007]. The effluent standard for boron is also 1 mg/L in Indonesia [Government of Indonesia, 2001].

1.3. Fluoride

Fluoride exists in mineral deposits and natural leaching process brings fluoride to groundwater. Anthropogenic activities are the source of fluoride pollution in surface waters

[Ayoob and Gupta, 2006]. Fluoride is widely used in industrial processes, in which important fluoride materials utilized in industry include fluorite (CaF_2) and hydrogen fluoride (HF) [WHO, 2002]. The wide applications of fluoride ion in industry result in wastewater with high concentration of fluoride from semiconductor manufacturing, metal processing, rubber and fertilizer production, electroplating, production of high-purity graphite, electrolysis of alumina, ceramic and glass manufacturing, and in nuclear application [Ochoa-Herrera et al., 2009; Wajima et al., 2009; Lee et al., 2010; Solangi et al., 2010]. WHO sets the drinking water standard of fluoride in the range of 0.5 – 1.0 mg/L [WHO, 2002]. The effluent standard for fluoride from industrial wastewater varies from country to country. For example, it is 5-15 mg/L in Japan [Yang et al., 1999], 15 mg/L in Taiwan [Hu et al., 2005; Chang and Liu, 2007], 25 mg/L in Poland [Grzmil and Wronkowski, 2006], 10 mg/L in China [Shen et al., 2003], 4 mg/L in the USA, and 2 mg/L in Iran [Khatibikamal et al., 2010].

1.4. Phosphate

Many industries use phosphorus in their manufacturing processes, such as food and dairy processing, metal treatment and etching, electronics, and fertilizer [Schipper et al., 2001]. The main source of phosphorus for industrial purpose is phosphate rock with annual consumption approaching 150 million tons [Abouzeid, 2008]. Special attention has been given to recycle of phosphorus since there is currently no known substitute [de-Bashan and Bashan, 2004].

Wastewater is the important source of phosphorus in the aquatic systems, mainly related to agricultural runoff, domestic and industrial wastewaters [Ann et.al, 2000]. Yapijakis and Wang [2006] indicated that total phosphorus ranged from 3,000 to 5,000 mg/L as P in retention pond of raw fertilizer wastewater. The fertilizer manufacturer also generates large amount of sludge known as phosphogypsum which may cause serious groundwater contamination of phosphate and fluoride if not properly disposed of [Battistoni et al., 2006]. Excess concentration of phosphorus will affect water quality, mainly through the depletion of the oxygen level as the algae decay [Morse et al., 1998; Georgantas and Grigoropoulou, 2007; Hosni et al., 2007]. The effluent standard of phosphate varies among different countries and is dependent on each industrial process. Regulation may also depend on receiving medium, such as land, surface water with different purpose, and coastal area; for example, in the USA, the effluent standard for phosphate from phosphate fertilizer manufacturer is 105 mg/L as P, and 35 mg/L as P from blended fertilizer [Yapijakis and Wang, 2006]. In Japan, the national effluent standard for phosphate is 16 mg/L with daily average of 8 mg/L [Japan MOE, 2007]. Effluent standard of South Korea for total phosphate is 8 mg/L [Kim et al., 2010].

1.5. Silica

Silica concentration in feed water of boiler or reverse osmosis (RO) needs to maintain below certain value to prevent silica scale [Sheikholeslami and Zhou, 2000]. Recently, silica has been used in many industrial production processes, such as semiconductor, pulp and paper, cement and concrete, laundry, and glass. Chemistry of silica is very complex as it exists in surface and ground water in different forms [Iler, 1979; Hamrouni and Dhahbi, 2001]:

Dissolved silica containing monomers, dimers and polymers of silicic acid.
Colloidal silica, which results from high polymerization of silicic acid containing particles, and three dimensions gel networks.
Reactive silica containing monomers and dimers forms that react with ammonium molybdate within 10 min. The other forms are referred to as "non-reactive" silica.

Concentration of silica in the aquatic environment ranges from 1 to 20 mg/L in surface water and from 7 to 45 mg/L in the groundwater. The reaction of SiO_2 in water can be shown as:

$$SiO_2 + 2\,H_2O \leftrightarrow H_4SiO_4\,(aq) \tag{1}$$

This reaction depends strongly on pressure and temperature. Iler [1979] mentioned that H_4SiO_4 (aq) was not only the species found in the solution, pyrosilicic acid ($H_6Si_2O_7$) dimers were also formed according to the following polymerization reaction:

$$2\,H_4SiO_4 \leftrightarrow H_6Si_2O_7 + H_2O \tag{2}$$

Presence of polysilicic acid is expected to increase with increasing total silica concentration and more abundant at higher pressures and temperatures [Iler, 1979; Zotov and Keppler, 2002]. Presence of H_4SiO_4 was dominant in temperature below 600°C and pressure of 6-8 kbar. On the other hand, when temperature and pressure increase both H_4SiO_4 and $H_6Si_2O_7$ dimers exist in solution. The polymerization rate of silica depends on pH and concentration of silica in the solution [Ning, 2003]. At neutral and alkaline pH the polymerization occurs very fast and extremely slowly at pH of 2 -3.

The presence of silica in industrial wastewater can cause some trouble. Zeng et al. [2007] mentioned that concentration of silica in paper making wastewater ranged from 50 to 200 mg/L as SiO_2 and would reduce the heat exchange efficiency. Silica is present in geothermal wastewater, and the prevention of its scaling is important [Gallup et al., 2003]. Silica is also generated from construction site because cement contains high concentration of silicate [Tsai et al., 2011]. Effluent standard for silica varies in different countries and is dependent on its manufacturing process. For example, the effluent standard for silica of paper making wastewater is 50 mg/L in USA, Canada and Finland [Thompson et al., 2001].

1.6. Other Anions

Selenium is found in high-sulfur coals, but there are no significant deposits of individual selenium. Selenium exists in the five oxidation states, namely –II, -I, 0, +IV, and +VI. Selenite (SeO_3^{2-}) and selenate (SeO_4^{2-}) are the two selenium ions most commonly found in surface water. Selenium is used in glass industry for coloring purpose. In the last 5 years, selenium also emerges as an important part in mammographic instrument as "amorphous selenium detector" [Kabir et al., 2006]. Commercial selenium is found in the generated anode slimes, produced as a byproduct from copper electrolytic refining [Haygarth, 1994].

Selenium is an essential trace nutrient for humans and vertebrates, but it becomes toxic when the concentration is high [Huang et al., 2007]. The US EPA advises a maximum contaminant level (MCL) of 0.050 mg/L for drinking water. For other water sources, "chronic criterion" for aquatic life is set at 0.005 mg/L for total selenium [Huang et al., 2007].

Extensive use of catalyst in industrial plants drives the production of several heavy metals containing wastewater, including molybdenum. It is also utilized in steel and lubricant production as additive. Excess molybdenum discharge into water can pose threat to aquatic water as it contributes to acute anemia. The permissible limit of molybdenum is 0.07 mg/L in Japan [Fujita et al., 2006].

2. Chemical Precipitation

Chemical precipitation is the rapid formation of a solid phase due to a physical or chemical change in the solution and this process also describes a complex phenomenon resulting from the formation of supersaturation condition [Blais et al., 2008]. Precipitation occurs in three different steps including: (1) nucleation, (2) crystal growth, and (3) flocculation. Nucleation starts with a supersaturated solution and under such conditions a condensation of ions will occur, forming small particle. This process is followed by growth of this tiny particle until the stage of supersaturation and the formation of the solid phase [Stumm and Morgan, 1996]. Precipitation is a complex mechanism that involves many parameters, including pH, degree of saturation, solubility product, ionic strength, temperature, and presence of interfering ions [Huang and Liu, 1999; Chuang et al., 2002; Aldaco et al., 2007; Chang and Liu, 2007]. They are delineated as following:

2.1. Solubility Product

Precipitate is formed only and if only when the concentration of each element ions in solution exceeds the solubility product, represented by Ksp. The general expression is:

$$A_aB_b\,(s) \rightarrow aA^{z+}(aq) + bB^{z-}(aq) \tag{3}$$

$$Ksp = [A^{z+}]^a [B^{z-}]^b \tag{4}$$

where $[A^{z+}]$ and $[B^{z-}]$ are the concentrations of ions in solution, and z^+ and z^- are the valences of the ions. It is noted that different solubility product of precipitate may be found. For example, the solubility product of calcium fluoride ranges from $10^{-10.5}$ to $10^{-10.6}$. Doyle and Parson [2002] indicated that the difference can be explained by different complexes formation as considered in an equilibrium and speciation considered in reaction. Solubility is also the key parameter in solution thermodynamics, and is related to the Gibbs free energy change:

$$\Delta G = \frac{-RT}{n}\ln\frac{IAP}{Ksp} \tag{5}$$

where R is the ideal gas constant, T is the absolute temperature, n is the number of ion in precipitate, IAP is free ionic activity product, and Ksp is the thermodynamic solubility product. The IAP value is determined by taking into account a series of chemical equilibrium characteristics of the system studied and it is related to saturation index (SI) value as follows:

$$SI = \log\frac{IAP}{Ksp}\text{, thus } \Delta G\text{: } \frac{-2.303RT}{n} SI \tag{6}$$

This equation can describe relation between ΔG and SI, as following:

when SI = 0, hence $\Delta G = 0$, the solution is in equilibrium,
when SI < 0, $\Delta G > 0$, the solution is undersaturated and precipitation is impossible in solution,
when SI > 0, $\Delta G < 0$, the solution is supersaturated and precipitation is spontaneous.

Song et.al. [2002] mentioned that SI is a good indicator to determine thermodynamic driving force for precipitation. However, from reaction kinetics point of view, supersaturation does not necessarily mean the quick, spontaneous precipitation because there is still a metastable zone, in which solution is already saturated but there is no precipitation over a long period [Stumm and Morgan, 1996].

2.2. Degree of Supersaturation

Supersaturation is the driving force for precipitation, by real (c) and equilibrium (c_o) solute concentration at a given pressure and temperature [Bristow et al., 2001]. The supersaturation links to the difference of chemical potential:

$$S = \frac{\Delta\mu}{k_B} = \frac{\ln\gamma_c}{\gamma_o c_o} \tag{7}$$

where $\Delta\mu$ is the difference of chemical potential, γ and γ_o are the corresponding activity coefficients, and k_B is Boltzmann constant. Bouropoulos and Koutsoukos [2000] showed that the degree of supersaturation can be defined as a ratio of the ion activity product to the solubility product. Take calcium fluoride as an example:

$$\Omega = \left(\frac{\alpha_{Ca^{2+}}(\alpha_{F^-})^2}{Ksp}\right) \tag{8}$$

where α represent ion activity of each ion species and Ksp is the solubility product of calcium fluoride. The effect of initial fluoride concentration on nucleation will be much larger than initial concentration of calcium in dilute fluoride-bearing solution [Yang et.al., 2011]. The nucleation occurs initially in precipitation processes as a result from supersaturation in solution. However, in metastable region the primary nucleation may not occur but secondary growth phenomena would be found [Forrest et al., 2008]. The secondary growth is known as a growth of the nuclei crystal through agglomeration with other particle. Crystallization process is favorable in this condition and produces large crystals. For example, large size of struvite crystal is formed with diameter > 1 mm under supersaturation ratio of 3 - 5 [Britton et al., 2005].

2.3. pH

pH is the most important parameters in precipitation process. The activity and speciation of each ion generally depends on pH condition. For example, struvite precipitation is dependent on pH because both activity of PO_4^- and NH_4^+ is determined by pH. In moderately low to neutral pH, the predominant species of arsenic are $H_2AsO_3^-$ and $HAsO_4^{2-}$, hence it will form precipitate when in contact with metal salts, such as iron [Pakzadeh and Batista, 2011]. Speciation diagram for As (V) and phosphate are similar as shown in Figure 1 and Figure 2. Both figures were based on thermodynamic model using WATEQ4F database.

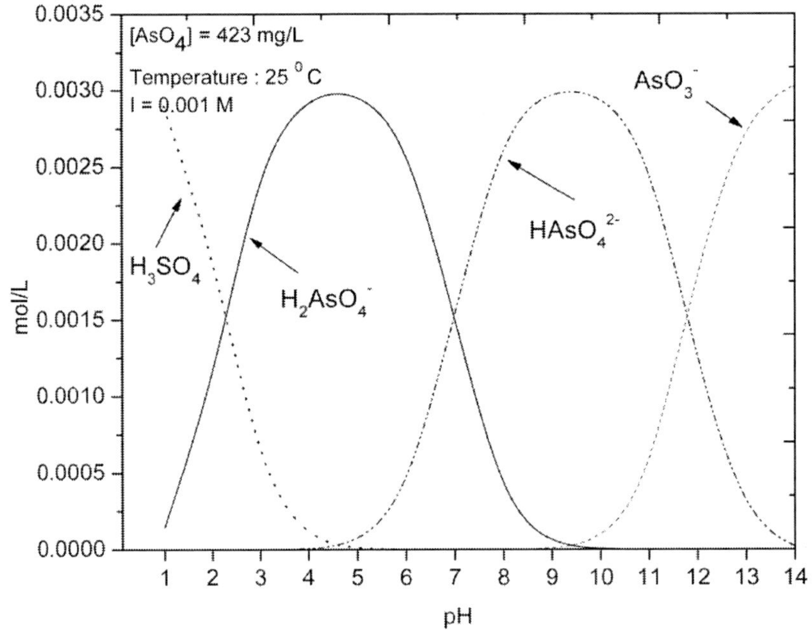

Figure 1. Speciation of arsenic (V) under different pH.

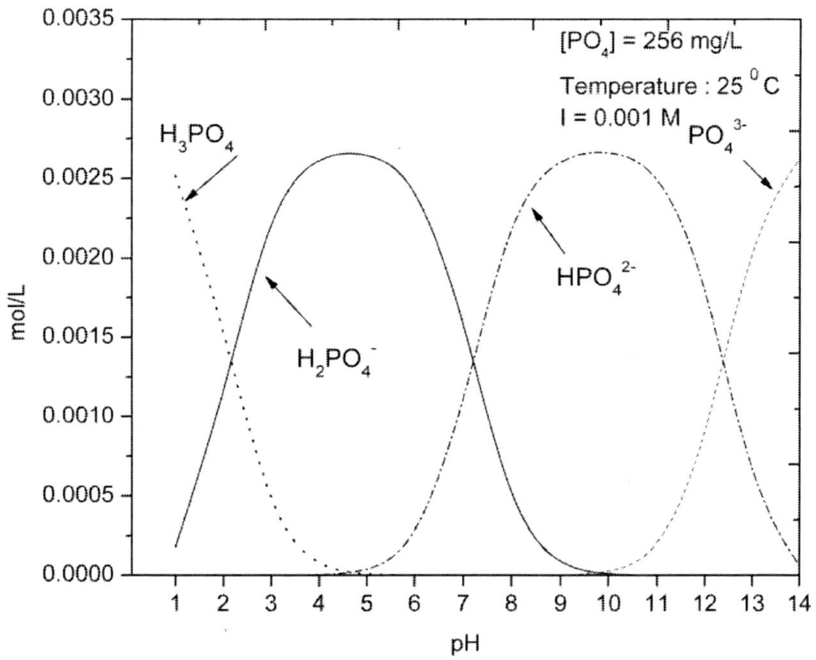

Figure 2. Speciation of phosphate under different pH.

Speciation of other anions, such as fluoride, borate, and selenium is also dependent on pH. pH also affects the effectiveness of precipitation reagents, such as CaO, $CaCl_2$, MgO, $Mg(OH)_2$, $MgCl_2$, and etc. Take precipitation of phosphate using magnesium salt as an example, magnesium compounds that can be used include MgO, $Mg(OH)_2$ and $MgCl_2$ [Burns et al., 2003]. There are many advantages in using MgO and $Mg(OH)_2$ as raw materials because of its cost and alkaline property [Chimenos et al., 2006]. However, MgO is favorable only for acidic wastewater because of its limited solubility. Warmadewanthi and Liu [2009a] found that $MgCl_2$ performed better in removal of phosphate from semiconductor wastewater under neutral and alkaline pH. It is in agreement with literature that $MgCl_2$ is more effective precipitation reagent for phosphate because of its high solubility under broad pH range [Uludag-Demirer et al., 2005]. pH also has significant impact on mechanism of interfering ions in precipitation process. Presence of carbonates significantly reduces the precipitation by competition between CO_3^{2-} and PO_4^{3-} for Ca^{2+} in pH range of 9-11 [Cao and Harris, 2008].

2.4. Ionic Strength

Ionic strength, I, determines the activity coefficient (γ_i) of each species in the aqueous solution. It can be expressed as the extended Debye-Hückel equation proposed by Davies [Parkhurst and Appelo, 1999]:

$$\log \gamma_i = Az_i^2 \left(\frac{\sqrt{I}}{1+\sqrt{I}} - 0.2\, I \right) \tag{9}$$

The thermodynamic approach can predict the effect of ionic strength in precipitation process. Many researchers pointed out that ionic strength affected the SI in precipitation [Song et al., 2002; Zhang et al., 2010]. Let us take calcium fluoride precipitation in the TFT-LCD wastewater as an example, in which initial fluoride concentration is 79.5 mg/L, molar ratio of [Ca]:[F] of 0.5:1, pH of 4, and at 25°C. Modeled results show that precipitation of CaF_2 decreases with increasing ionic strength (Table 1).

Table 1. Effect of ionic strength on CaF_2 precipitate

Ionic strength (M)	SI value of CaF_2	Precipitate (M)
0.001	2.817	1.86E-03
0.005	2.722	1.85E-03
0.01	2.657	1.83E-03
0.05	2.431	1.78E-03
0.1	2.310	1.75E-03
0.5	2.161	1.70E-03

2.5. Temperature

The association or dissociation reactions of chemical species in solution and thermodynamic solubility are all temperature dependent [Song et al., 2002]. The ion activity and the SI value depend on temperature as described in the equation of the Gibbs free energy change (Equation **Error! Reference source not found.**). Again, let us take calcium fluoride precipitation in the TFT-LCD wastewater as an example, the thermodynamic model prediction shows that increasing temperature decreases SI value and increases presence of fluoride ion in solution at pH of 4 (Table 2).

Table 2. At pH 4, the solubility of fluoride decreases with increasing temperature and the concentration of HF increases significantly. However, as pH increases to 6.5 temperature no longer has significant effect on precipitation reaction

T (°C)	Specification (M)				CaF_2 (M)		SI	
pH	4		6.5		4	6.5	4	6.5
	HF	F^-	HF	F^-				
5	4.26E-05	4.14E-04	1.39E-07	4.24E-04	1.87E-03	1.89E-03	2.77	2.85
10	4.62E-05	4.14E-04	1.51E-07	4.25E-04	1.87E-03	1.89E-03	2.76	2.84
20	5.46E-05	4.12E-04	1.79E-07	4.25E-04	1.87E-03	1.89E-03	2.73	2.84
25	5.95E-05	4.11E-04	1.96E-07	4.25E-04	1.87E-03	1.89E-03	2.72	2.84
30	6.50E-05	4.10E-04	2.36E-07	4.25E-04	1.86E-03	1.89E-03	2.71	2.83
50	9.34E-05	4.04E-04	3.15E-07	4.26E-04	1.85E-03	1.89E-03	2.64	2.81
70	1.36E-04	3.96E-04	4.72E-07	4.28E-04	1.83E-03	1.89E-03	2.55	2.78
90	2.01E-04	3.84E-04	7.22E-07	4.29E-04	1.81E-03	1.89E-03	2.43	2.76

2.6. Interfering Ions

Generally, the presence of co-existing ion will affect precipitation reaction with different mechanism involved. Phosphate hindered arsenic precipitation because both ions have similar properties [Li et al., 2010]. The decrease in precipitation efficiency is mostly due to competition between anions for precipitation reagent, such as competition between carbonate and phosphate [Cao and Harris, 2008], or between fluoride and phosphate [Grzmil and Wronkowski, 2006]. As mentioned, the competition depends on pH and concentration of each ions involved in the precipitation process. On the other hand, presence of other ions might decrease or increase the induction time for precipitation or crystallization. Presence of high concentration of sodium reduced induction time of struvite precipitation, while the presence of sulfate increased induction time of struvite precipitation [Kabdasli et al., 2006].

3. Chemical Precipitation of Anions

3.1. Chemical Precipitation of Phosphate

The most suitable precipitation reagents for chemical precipitation of phosphorus are calcium, iron and aluminum [Jiang and Graham, 1998; Morse et al., 1998; Maurer and Boller, 1999; Penetra et al., 1999]. Calcium precipitation is a common method for phosphorus removal, mainly because of low cost and ease of handling [Yi and Lo, 2003]. This precipitation reaction is an important process in controlling P activity and availability in environmental system [Hu et al., 2005]. Removal of phosphate is achieved by direct precipitation of calcium phosphate known as hydroxyapatite ($Ca_{10}(PO_4)_6(OH)_2$, HAP).

$$10\,Ca^{2+} + 6\,PO_4^{3-} + 2\,OH^- \rightarrow Ca_{10}(PO_4)_6(OH)_2 \tag{10}$$

This mineral is the most stable form of calcium phosphorus precipitate based on thermodynamic approach [Abbona et al., 1988; Jiang and Graham, 1998]. Although HAP is a very stable phase of calcium phosphate precipitate, other aspects such as the reaction kinetics need to be considered as well. Musvoto et al. [2000] showed the initial precipitate formed is either brushite ($CaHPO_4.2H_2O$) (DCPD) or octacalcium phosphate ($Ca_4H(PO_4)_3$) (OCP), which is then slowly changed into HAP. Several different chemical reactions of calcium and phosphate are:

$$Ca^{2+} + HPO_4^{2-} \rightarrow CaHPO_4 \text{ (monetite) (DCPA)} \tag{11}$$

$$Ca^{2+} + H^+ + PO_4^{3-} + 2\,H_2O \rightarrow CaHPO_4.2\,H_2O \tag{12}$$

$$4\,Ca^{2+} + 3\,PO_4^{3-} + H^+ \rightarrow Ca_4H(PO_4)_3 \tag{13}$$

Iron metal salts, such as $FeSO_4$ and $FeCl_2$ are effective precipitation reagents to remove phosphate as strengite ($FePO_4.2H_2O$) [Fytianos et al., 1998; Grzmil and Wronkowski, 2006; Zhang et al., 2010]. The chemical reaction of strengite is:

$$Fe^{3+} + PO_4^{3-} + 2\,H_2O \rightarrow FePO_4.2H_2O \qquad (14)$$

Fytianos et al. [1998] showed that chemical reactions are very complex and result in a much larger iron demand than that required by the precipitation reaction alone. The most important parameters affecting the extent of phosphate removal are iron dose, final pH and initial phosphate concentration [Pierri et al., 2000]. Aluminium hydroxide ($Al(OH)_3$) and alum also are effective precipitation reagents for phosphorus treatment process with chemical reaction:

$$2\,Al^{3+} + 2\,PO_4^{3-} \rightarrow 2\,AlPO_4 \qquad (15)$$

Compared with calcium and iron, aluminum is more expensive but is very effective [Fytianos et al., 1998; Zhang et al., 2010].

Magnesium salts can be a good alternative in recovering phosphate from wastewater, knowing that they have very low solubility and the precipitates can be reused for many industrial applications [Chimenos et al., 2006]. Reactions of magnesium with phosphate form three possible magnesium phosphate precipitates; namely, magnesium hydrogen phosphate or newberryite ($MgHPO_4.3H_2O$), trimagnesium phosphate ($Mg_3(PO_4)_2.22H_2O$), and bobierrite ($Mg_3(PO_4)_2.8H_2O$) [Musvoto et al., 2000].

$$Mg^{2+} + HPO_4^{2-} + 3\,H_2O \rightarrow MgHPO_4.3\,H_2O \qquad (16)$$

$$3\,Mg^{2+} + 2\,PO_4^{3-} + 22\,H_2O \rightarrow Mg_3(PO_4)_2.22\,H_2O \qquad (17)$$

$$3\,Mg^{2+} + 2\,PO_4^{3-} + 8\,H_2O \rightarrow Mg_3(PO_4)_2.8\,H_2O \qquad (18)$$

The other possibility for magnesium phosphate recovery is reaction at appropriate concentrations of magnesium, phosphate, and ammonium to form struvite ($MgNH_4PO_4.6H_2O$):

$$Mg^{2+} + NH_4^{+} + PO_4^{3-} + 6\,H_2O \rightarrow MgNH_4PO_4.6\,H_2O \qquad (19)$$

This process has been extensively studied to remove phosphate and ammonium from various different sources of wastewater. Different solubility product for calcium phosphate precipitation and magnesium phosphate precipitation are listed in Table 3.

The precipitation of calcium phosphate is affected by the molar ratio of [Ca]:[P]. Hosni et al. [2007] showed that HAP is found at molar ratio of 1.67 and tricalcium phosphate ($Ca_3(PO_4)_2$) is found at molar ratio of 1.5. Excess concentration of calcium in solution will enhance removal of phosphate from wastewater [Moriyama et al., 2001; Yi and Lo, 2003; Liu et al., 2009]. Similarly, for strengite, increasing concentration of iron will increase removal efficiency [Fytianos et al., 1998; Thistleton et al., 2002].

Table 3. Solubility product of metal phosphate precipitation

Mineral phase	Reaction	log K at 25°C
DCPA	$Ca^{2+} + HPO_4^{2-} \leftrightarrow CaHPO_4$	-6.6 [τ]
	$Ca^{2+} + H^+ + PO_4^{3-} \leftrightarrow CaHPO_4$	-19.275 [β]
	$Ca^{2+} + H_2PO_4^- \leftrightarrow CaHPO_4 + H^+$	0.3 [δ]
DCPD	$Ca^{2+} + H^+ + PO_4^{3-} + 2H_2O \leftrightarrow CaHPO_4.2H_2O$	-18.995 [β]
	$Ca^{2+} + HPO_4^{2-} + 2H_2O \leftrightarrow CaHPO_4.2H_2O$	-25.2 [ζ]
	$Ca^{2+} + H_2PO_4^- + 2H_2O \leftrightarrow CaHPO_4.2H_2O + H^+$	0.629 [δ]
OCP	$4Ca^{2+} + 3PO_4^{3-} + H^+ \leftrightarrow Ca_4H(PO_4)_3$	-46.9 [τ]
	$4Ca^{2+} + 3PO_4^{3-} + H^+ + 3H_2O \leftrightarrow Ca_4H(PO_4)_3.3H_2O$	-47.95 [β]
	$4Ca^{2+} + 3PO_4^{3-} + H^+ + 3H_2O \leftrightarrow Ca_4H(PO_4)_3{:}3H_2O$	-48 [ζ]
	$4Ca^{2+} + 3H_2PO_4^- + 3H_2O \leftrightarrow Ca_4H(PO_4)_3{:}3H_2O + 5H^+$	11.75 [δ]
HAP	$10Ca^{2+} + 6PO_4^{3-} + 2OH^- \leftrightarrow Ca_{10}(PO_4)_6(OH)_2$	-114 [τ]
	$5Ca^{2+} + 3PO_4^{3-} + H_2O \leftrightarrow Ca_5(PO_4)_3(OH) + H^+$	-44.333 [β]
	$5Ca^{2+} + 3HPO_4^{2-} + H_2O \leftrightarrow Ca_5(PO4)_3(OH) + 4H^+$	-3.421 [η]
	$5Ca^{2+} + 3PO_4^{3-} + OH^- \leftrightarrow Ca_5(PO4)_3(OH)$	-58.5 [ζ]
	$10Ca^{2+} + 6H_2PO_4^- + 2H_2O \leftrightarrow Ca_{10}(PO_4)_6(OH)_2 + 14H^+$	28.9 [δ]
Struvite	$Mg^{2+} + NH_4^+ + PO_4^{3-} + 6H_2O \leftrightarrow MgNH_4PO_4.6H_2O$	-13.15[ε]
	$Mg^{2+} + NH_4^+ + PO_4^{3-} + 6H_2O \leftrightarrow MgNH_4PO_4.6H_2O$	-12.6[τ]
	$Mg^{2+} + NH_4^+ + PO_4^{3-} + 6H_2O \leftrightarrow MgNH_4PO_4.6H_2O$	-13.26[φ]
Newberryite	$Mg^{2+} + HPO4^{2-} + 3H_2O \leftrightarrow MgHPO_4.3H_2O$	-5.8[φ]
Bobierrite	$3Mg^{2+} + 2PO_4^{3-} + 8H_2O \leftrightarrow Mg_3(PO_4)_2.8H_2O$	-25.2[φ]
Trimagnesium phosphate	$3Mg^{2+} + 2PO_4^{3-} + 22H_2O \leftrightarrow Mg_3(PO_4)_2.22H_2O$	-23.1[φ]
Strengite	$Fe^{3+} + PO_4^{3-} + 2H_2O \leftrightarrow FePO_4.2H_2O$	-12.6[ψ]
$AlPO_4$	$2\,Al^{3+} + 2\,PO_4^{3-} \leftrightarrow 2AlPO_4$	-20[τ]

[τ]: Stumm and Morgan, 1996
[β]: Parkhurst and Appelo, 1999
[η]: Gustafsson, 2005
[ζ]: Strang and Wareham, 2006
[δ]: Phelan and Mattigod, 1987
[ε]: Çelen et al., 2007
[φ]: Musvoto et al., 2000
[ψ]: Zhang et al., 2010

Based on computer model, concentration of ferric ion salts was the most limiting constituent for strengite [Zhang et al., 2010]. Aluminum and magnesium are needed in excess concentration to remove phosphate from wastewater [Bouropoulos and Koutsoukos, 2000; Münch and Barr, 2001; Stratful et al., 2001; Özacar and Sengil, 2003; Uludag-Demirer et al., 2005; Wang et al., 2005; Warmadewanthi and Liu, 2009a]. Stratful et al. [2001] claimed that precipitation of struvite will be limited by ammonium ion (NH_4^+). Theoretically, only 85%

nitrogen exists as ammonium ion in pH range of 9 - 10, and excess NH_4^+ is necessary to increase removal efficiency [Stratful et al., 2004; Wang et al., 2005]. Warmadewanthi and Liu [2009b] also found the similar result in precipitation of struvite from semiconductor wastewater.

Because of different speciation as affected by pH, different mechanism of precipitation will occur under different pH. The optimum pH for precipitation of calcium phosphate is under alkaline pH, and different precipitate might be found in solid phase in different pH condition [House, 1999; Tchobanoglous et al., 2003; Liu et al., 2009]. Though HAP is the predominant solid, precipitation of OCP appears to be limited in narrow range pH of 6 – 7, and DCPD was formed at pH > 7 [De Rooij et al., 1984]. van-Kemenade and Bruyen [1987] found that above pH 6.7 the order of increasing stability of calcium phosphate precipitation is HAP > DCPD > OCP, and the stability sequence changes to HAP > OCP > DCPD at pH 6.7. Experimental results of iron phosphate precipitation showed that the removal of phosphate is optimum in pH range of 4 - 5.5 at [Fe]:[P] stoichiometric ratio of 1:1 [Pierri et al., 2000; Thistleton et al., 2002]. Banu et al. [2008] used alum to remove phosphate from low alkalinity wastewater and the optimum pH was 5.7 - 5.9. Addition of coagulant aids, such as polyelectrolyte and clay can increase removal efficiency and lower optimum dosage for alum phosphate precipitation [Özacar and Sengil, 2003]. For magnesium phosphate precipitate, bobierrite was stable in pH range of 8 to 10 [Warmadewanthi and Liu, 2009a]. Newberryite was found at pH < 7 and with high excess concentration of magnesium [Babic-Ivancic et al., 2006] and trimagnesium phosphate tends to form at pH 9 and above [Le Corre et al., 2007]. However, the reaction rate of bobierrite was slow [Mamais et.al., 1994; Chimenos et al., 2006]. It has been well documented that optimum pH for struvite precipitation is from 8.0 to 10 [Ohlinger et al., 1998; Musvoto et al., 2000; ; Stratful et al., 2001; Chimenos et al., 2003; Nelson et al., 2003]. This optimum pH range is related to solubility of PO_4 and the presence of nitrogen as ammonium (NH_4^+) ion in solution [Doyle and Parsons, 2002; Levlin and Hultman, 2003].

Regarding the presence of co-existing ion, carbonate (CO_3^{2-}) significantly decreased removal efficiency of phosphate in range pH of 9 – 11 [Battistoni et al., 1997; Kapolos and Koutsoukos, 1999; Song et al., 2002; Montastruc et al., 2003;Cao and Harris, 2008]. It is due to the competition between CO_3^{2-} and PO_4^{3-} for Ca^{2+}. van der Houwen and Vaisami-Jones [2001] mentioned that the presence of citrate in solution also inhibited HAP precipitation by forming phosphate citrate complex. Cao et al. [2007] studied the impact of presence of Mg^{2+}, SO_4^{2-}, CO_3^{2-}, two organic acids (oxalic and humic acid), and two Si-rich solids to calcium phosphate precipitate. Result revealed that Mg^{2+}, humic acid, and oxalic acid significantly inhibited HAP phase. On the other hand, sulphate and Si-rich solid had less impact on precipitation of HAP except reducing HAP precipitation in the early stage. Grzmil and Wronkowski [2006] showed that the presence of fluoride in wastewater will hinder precipitation of $FePO_4$ due to the formation of iron fluoride complex. Again, fluoride also hinders precipitation of $AlPO_4$ because aluminum salts will form highly stable complex with fluoride [Grzmil and Wronkowski, 2006]. Tannic acid and other organic material inhibited the alum-phosphate precipitate [Omoike and vanLoon, 1999].

Presence of other ions may also affect precipitation of magnesium phosphate. The presence of calcium ion will affect struvite precipitation because of competition between Mg and Ca for phosphorus [Koutsoukos et al., 2003; Wang et al., 2005]. Le Corre et al. [2005] also found that at high molar ratio of calcium to magnesium struvite precipitation was hindered. However, the interference can be minimized by adjusting molar ratio of magnesium to calcium in solution and efficient struvite formation can be achieved [Battistoni et al., 2000; Wang et al., 2005]. Other major ions that are commonly found in wastewater, such as carbonate-bicarbonate, sodium, and sulfate, may also affect struvite precipitation [Kabdasli et al., 2006]. Carbonate can form precipitate with magnesium and has a slight effect on induction time of struvite formation. Warmadewanthi and Liu [2009a; 2009b] found that the presence of interfering ions, such as fluoride, chloride, and sulfate did not affect struvite and bobierrite precipitation. Although Mg may react with F to form MgF_2 precipitate, however, magnesium preferentially formed magnesium phosphates precipitate under alkaline conditions [Battistoni et al., 2006; Warmadewanthi and Liu, 2009a]. Kim et al. [2009] claimed that mixing energy has a significant effect in avoiding inhibition effect of fluoride on struvite precipitation.

Precipitation of calcium phosphate decreases significantly with gradual increase of ionic strength. Removal efficiency decreases because of decreasing activity coefficient of phosphate ion under higher ionic strength [Liu et al., 2009]. Zhang et al. [2010] used computer model to show the effect of ionic strength on $FePO_4.2H_2O$. The increasing ionic strength will decrease the SI of $FePO_4.2H_2O$. For struvite precipitation, many researchers reported that the presence of complexing reagents formed complex with Mg^{2+} under higher ionic strength. This will decrease the magnesium availability to form struvite precipitate [Ohlinger et al., 1998; Bouropoulos and Koutsoukos, 2000; Uludag-Demirer et al., 2005].

As mentioned, the SI of precipitate is affected by temperature of solution. van Kemenade and de Bruyn [1987] already showed the stability of calcium phosphate precipitate under different temperature. Song et al., [2002] used computer model to predict precipitation of HAP and found that increasing temperature will increase SI linearly. Zhang et al. [2010] also assessed effect of temperature on $FePO_4.2H_2O$ precipitation. The optimum temperature is 25 – 30°C. The lowest pKsp of stuvite was found at 30°C and will increase with increasing temperature under range 30-60°C [Bhuiyan et al., 2006; Hanhoun et al., 2011].

Utilization of Phosphate Precipitate

Recovery of phosphate precipitates has been a great concern. Special attention is given to recycle of phosphorus as a raw material for the phosphorus industry [de-Bashan and Bashan, 2004]. Regarding separation of solid from solution, sedimentation is generally used, usually following polymer flocculation [Tchobanoglous et al., 2003]. Different methods have been examined for separation, such as membrane filtration and flotation processes. A hybrid flotation and membrane microfiltration could remove phosphate from solution [Peleka et al., 2006]. Liu et al. [2009] showed that precipitation-flotation can effectively separate phosphate precipitate from solution. A hybrid precipitation-microfiltration process showed higher removal efficiency of phosphate and better solid separation from TFT-LCD wastewater [Lu and Liu, 2010]. Advantages of these processes include flexible operation, small footprint, effective separation, and potential reuse of sludge material.

Chimenos et al. [2006] showed that low grade of MgO can be used to remove and recover phosphate. Newberryite and bobierrite precipitates from this precipitation process can be reused in phosphorus industry. Warmadewanthi and Liu [2009a] proposed the formation of bobierrite in solid phase from semiconductor wastewater with reaction time of 40 minute by using $MgCl_2$ to induce precipitation. This precipitate is easier to remove from solution by sedimentation and can potentially be reused.

Chemical precipitation of struvite is the most promising method to recover phosphate simultaneously with ammonia in wastewater [de-Bashan and Bashan, 2004]. Struvite has advantages of slower release rate, less frequent application required, no fertilizer burn problem even at high application rate, and less impurities as compared with other commercial fertilizers [Münch and Barr, 2001; Wang et al., 2005]. The other possible application of struvite is as raw material in production of detergent, cosmetics, and animal feed [Gaterell et al., 2000]. Various methods have been examined to recover struvite from different kinds of wastewater. For example, recovery of struvite crystal from sewage wastewater [Jaffer et al., 2002], urine [Udert et al., 2003], anaerobic supernatant and swine manure [Yoshino et al., 2003; Battistoni et al., 2005; Suzuki et al., 2007; Türker and Çelen, 2007], landfill leachate [He et al., 2007; Kim et al., 2007], agro industry waste [Moerman et al., 2009], semiconductor wastewater [Kim et al., 2009; Warmadewanthi and Liu, 2009b], up-flow sludge blanket (UASB) reactor [Yetilmezsoy and Sapci-Zengin, 2009]. Recently, pH of struvite precipitation can be controlled by CO_2 bubbling [Korchef et al., 2011; Rahman et al., 2011]. The full scale fluidized bed reactor was also developed in Japan for recovery of struvite. Mineral was sold as a fertilizer at price of approximately USD $250 per tons [Shimamura et.al., 2003]. In Canada, two technical scale fluidized bed reactors had been proposed [Forrest et al., 2008]. From economic point of view, recovery of struvite from 100 m^3/day of digester supernatant can reduce sludge treatment cost of approximately AUD 1,333/day and recovery of this element around the world will reduce phosphate rock mining 1.6% annually [Shu et al., 2006].

3.2. Chemical Precipitation of Arsenic

Precipitation facilitates arsenic removal by converting ionic arsenic into particulate arsenic, which then can be separated using sedimentation or flotation. Numerous studies on arsenic precipitation have been conducted, mostly focusing on As (V). Precipitation using lime or ferrous compounds, which result in the formation of calcium arsenate or ferric arsenate, is among the most common method investigated intensively and is found to be effective [Hering et.al., 1996; Bothe Jr. and Brown, 1999; Pakzadeh and Batista, 2011;]. The process itself also depends on several factors, including arsenic speciation, precipitation reagents, pH, and interfering ions.

Since precipitation relies on the reaction between arsenic and precipitation reagent, species of soluble arsenic is an important factor to be considered. For example, As (V) is dominant in its ionic form at pH >3, namely, $H_2AsO_4^-$, $HASO_4^{2-}$, and AsO_4^{3-} (see Figure 1), while As (III) is neutral al pH less than 9, as depicted in Figure 3, based on thermodynamic model using WATEQ4F database.

The zero charge of dominant As (III) species in neutral pH often contributes to the difficulty in removing arsenic, as it is also noted in adsorption process. Major species of As (V), however, carry -1 or -2 charge in neutral pH so the efficiency of arsenic removal is higher. Hence, As (III) often needs to be oxidized before precipitation [Wang et al., 2011].

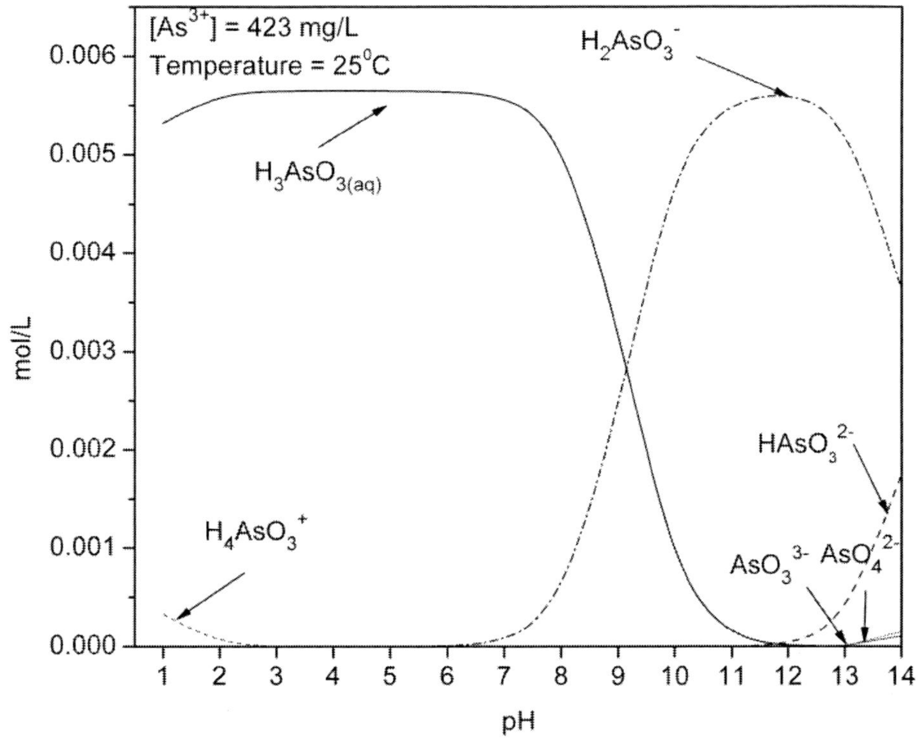

Figure 3. Speciation of As (III) under different pH.

Arsenic can be precipitated using metal compounds, such as ferric, calcium, barium, magnesium, and strontium. It can also be precipitated as sulphide. Calcium is widely used to treat arsenic containing wastewater. Several forms of calcium arsenates might be produced, either hydrate or anhydrate. Hydrate form, such as $Ca_3(AsO_4)_2.xH_2O$ is often considered thermodynamically stable under aqueous conditions when compared with anhydrate $Ca_3(AsO_4)_2$ or $Ca_5(AsO_4)_3OH$. Molar ratio of Ca and As is typically maintained at higher than 1.5. The anhydrate $Ca_5(AsO_4)_3OH$ is often referred as arsenate apatite, and having solubility of less than 0.2 mg As/100 L [Bothe Jr. and Brown, 1999]. The reaction of arsenate apatite formation is written as follows:

$$3\ AsO_4^{3-} + 5\ Ca(OH)_2 \rightarrow Ca_5(AsO_4)_3(OH) + 9\ OH^- \qquad (20)$$

The stability of arsenate apatite is lower compared to hydrate calcium arsenates; however stable phase can be obtained via several operating conditions, for instance in the absence of magnesium [Bothe Jr. and Brown, 1999]. Stable arsenate apatite was also formed at temperature higher than 100°C [Itakura et al., 2007b].

Ferrous compounds are also broadly employed in arsenic removal. The ferrous compounds include hydrate or anhydrate ferric chloride ($FeCl_3$), ferric sulfate ($Fe_2(SO_4)_3$), or ferrous sulfate ($FeSO_4$). A study reported that compounds formed from arsenic precipitation using ferrous sulfate were scorodite ($FeAsO_4.2H_2O$), basic iron sulfate ($FeOHSO_4$), "Type 1" ($Fe_2(HAsO_4)_3.zH2O$ with z<4, and "Type 2" ($Fe_4(AsO_4)_3(OH)_x(SO_4)_y$ with x+2y=3) at temperature above ambient conditions, 150° to 225°C, and Fe/As ratio of 1/1 to 9/1 [Swash and Monhemius, 1994]. Precipitation at high temperature, referred as hydrothermal mineralization, is commonly applied since it imitates mineral precipitation in nature, in which precipitation is caused by magma activity. Addition of $FeCl_3$ to bio-oxidized arsenate-containing solution also removed arsenate with efficiency higher than 95% [Andrianisa et al., 2008]. High arsenic removal was achieved at low pH, and decreased rapidly when pH increased [Hering et al., 1996]. In a moderately low to neutral pH, the predominant species are $H_2AsO_3^-$ and $HAsO_4^{2-}$, thus they will form precipitate when in contact with ferric ions. When pH is extremely low, non-ionic arsenic acid presents and there is no ferric oxides formed, both factors contribute to the null removal of arsenic [Pakzadeh and Batista, 2011]. Magnesium chloride ($MgCl_2$) and magnesium sulphate ($MgSO_4$) are also effective precipitation reagents. The precipitate resulted is $Mg_3(AsO_4)_2$, and optimum removal was achieved at pH between 8 and 10, and Mg to As molar ratio of at least 2:1 [Park et al., 2010].

The optimum pH for arsenic precipitation will be the pH at which arsenic species is least soluble, and it is significantly affected by solution properties. Precipitates formed have their own stability over certain pH. For instance, scorodite is stable at low pH only [Robins et al., 1987]. Magnesium arsenates, on the contrary, are stable at high pH [Park et al., 2010]; along with arsenate apatite [Bothe Jr. and Brown, 1999]. Apart from stability, the effect of pH on precipitation reagent is also important. Disadvantage of using magnesium compounds to precipitate arsenic is that they cannot be utilized at pH over 12 since magnesium will precipitate into magnesium hydroxide.

The presence of co-existing ions also affects arsenate precipitation. Organic matter in solution, such as humic acid, induces negative effect on As (V) precipitation using ferric salts. Less than 10% removal was obtained when humic acid existed in the solution as compared with 94% removal when in the absence of organic matter [Pallier et al., 2011]. A significant decrease of arsenic removal was also observed when phosphate was present, since the chemical properties of phosphate and arsenate are similar, while carbonates effect was less significant [Li et al., 2010].

Arsenic precipitates are usually very fine in size, sometimes even below 1 μm, and it is important to develop better solid/liquid separation processes. Following precipitation, microfiltration or adsorption is used. Addition of coarse calcite contributed to a significant increase in settling rate of arsenic precipitates from water [Song et al., 2006]. Ferric and manganese binary oxide (FMBO) adsorption also demonstrated an excellent performance, removing more than 99% of residual arsenic after precipitation process [Wang et al., 2011].

Utilization of Arsenic Precipitate

The arsenic sludge is commonly disposed of in landfill. Pre-drying of the sludge is needed since arsenic precipitates have high water content. In Bangladesh, such sludge is being dumped to cow dung bed [Ali et al., 2003]. Toxicity characteristic leaching procedure (TCLP) is performed prior to landfill to ensure the stability of arsenic precipitates. In India (and also Bangladesh), sludge with waste constituent such as arsenic, lead, and cadmium have also been mixed with cement block and plinths, however, drying and carbonation affected the release of those constituents, making it exposed to weathering [Sanchez et al., 2002].

3.3. Chemical Precipitation of Fluoride

Calcium salts, such as lime (CaO), $Ca(OH)_2$, and calcium chloride ($CaCl_2$) are generally used to precipitate fluoride from water [Reardon and Wang, 2000]. The stoichiometric reaction of calcium and fluoride to form fluorite (CaF_2) is:

$$Ca^{2+} + 2\,F^- \rightarrow CaF_2 \qquad (21)$$

Other metal salt that can be used for precipitation of fluoride is aluminum salt to form AlF_3 precipitate. The reaction is pH dependent in using Al_2O_3 according to the following chemical reaction [Aldaco et al., 2006]:

$$Al_2O_3 + 6\,HF \rightarrow 2\,AlF_3 + 3\,H_2O \qquad (22)$$

Alum is also utilized as chemical amendment for fluoride removal. However, the reactions tend to be a combination of precipitation and adsorption of fluoride on the surface of $Al(OH)_3$. Toyoda and Taira [2000] proposed this method to reduce sludge produce and running cost in removal of fluoride. Under certain conditions, aluminum has a higher bonding strength with fluoride compared with calcium. However, aluminum is much more expensive than the calcium salts, and its sludge has poor dewaterability. For these reasons, aluminum is not commonly used for concentrated fluoride wastewater.

Excess concentration of calcium is necessary; when in the absence of seeding materials to induce calcium fluoride precipitation [Stumm and Morgan, 1996]. It has been pointed out that pH has a significant effect on calcium fluoride precipitation. Yang et al. [2001] found that at pH > 4, there is no significant effect for precipitation of fluoride from electronics wastewater. Similar result was also found by Chang and Liu [2007], in which increasing residual fluoride in solution at pH < 5 was due to incomplete hydrolysis of HF, while at pH > 12 it was because of the formation of calcium hydroxide precipitate ($Ca(OH)_2$).

Warmadewanthi and Liu [2009c] found that the presence of PO_4 has significant effect on precipitation of fluoride under alkaline pH. Phosphate will compete with fluoride for calcium under pH range of 8 - 9 and residual fluoride in solution increases significantly at pH > 10 because of the formation of $Ca(OH)_2$. Grzmil and Wronkowski [2006] also mentioned that phosphate competes with fluoride for calcium and higher dosage of calcium is needed to effectively remove fluoride. Yang et al. [1999] and Urbansky [2002] found that silica can react with fluoride to form a complex as SiF_6^{-2}. At high molar ratio, $[SiO_2]$:[F], it delays precipitation of CaF_2. However, presence of colloidal silicate in chemical mechanical polishing (CMP) can enhance removal of fluoride under pH of 6.5 to 7 [de Luna et al., 2009]. Huang and Liu [1999] found that the presence of sulfate in certain concentration depressed fluoride removal efficiency because of the competition between sulfate and fluoride in solution. Also, the residual fluoride concentration increased significantly as ionic strength increases from 0.05M to 0.5 M $NaNO_3$.

Utilization of CaF_2 Precipitate

The potential recovery of CaF_2 from fluoride precipitation is of great interests [Aldaco et al., 2005]. To decrease amount of sludge generated, $CaCl_2$ may be a better choice over lime

[Huang and Liu, 1999]. Turner et al. [2005] and Aldaco et al. [2007] indicated that both precipitation and surface adsorption were found when using calcite for fluoride removal from solution. The fluidized-bed crystallization process has been demonstrated effective for simultaneous wastewater treatment and recovery of CaF_2 sludge as fluorite [van den Broeck, 2003]. The chemistry of the process is similar to traditional process except that its concentration, hydrodynamics, and seeding are controlled. It can also be conducted by adding aluminum and sodium to the fluidized-bed reactor to form cryolite ($NaAlF_6$) [Aldaco et al., 2005]. Silica sand could be used as the seeding materials [Aldaco et al., 2006]. Tai et al. [2006] investigated effect of metastable region for crystallization of CaF_2 and found that pH, saturation, and seed size affect the process. The pellets or grains produced from the fluidized-bed reactor contained 97% calcium fluoride [Aldaco et al., 2006; 2007]. Due to composition, crystal structure, and lower water content, this pellet can be used as a raw material in various applications. This process has been established in full scale among several semiconductor manufacturers in Taiwan.

3.4. Chemical Precipitation of Silica

Many methods have already been developed to remove silica from different kinds of water and wastewater. Generally, the chemical treatment process consists of precipitation, coagulation and flocculation [Sheikholeslami and Zhou, 2000]. Precipitation of silica may occur in presence of calcium, magnesium, and polyvalent metal ions, such as iron and aluminum [Bremere et al., 2000]. Calcium oxide reacts with silica, ranging from 580 mg/L to 900 mg/L, to form precipitate in wastewater from geothermal brine [Ueda et al., 2003]:

$$CaO + H_2O \leftrightarrow Ca(OH)_2 \tag{23}$$

$$Ca(OH)_2 \leftrightarrow Ca^{2+} + 2\ OH^- \tag{24}$$

$$Ca^{2+} + SiO_2(aq) + H_2O \leftrightarrow CaSiO_3 + 2\ H^+ \tag{25}$$

The reaction is dependent on initial concentration of silica, concentration of calcium, and pH in solution, as SiO_2 in aqueous phase is the dominant species under pH range of 7 to 10. This reaction is also affected by molar ratio of Ca to Si, and more precipitation occurs at ratio greater than 1. Gallup et al. [2003] found similar result in removing silica from water using lime or caustic-soda, and rapid flocculation and sedimentation of colloidal silica was observed at pH at 8.5 – 9.0. Magnesium hydroxide is commonly used to remove soluble and colloidal silica from water [Iler, 1979; Sheikholeslami and Bright, 2002]. Magnesium hardness is more effective than calcium hardness in increasing the rate of polymerization and reducing the solubility of silica in solution [Sheikholeslami et al., 2001]. Also, it has been indicated that $MgCl_2$ is more effective for silica removal than MgO [Zeng et al., 2007]. Adsorption reaction, chemical reaction, or a combination of both can be responsible for silica removal using magnesium salts [Sheikholeslami and Bright, 2002]. Chemical reactions involved in silica removal are shown as follows [Chen et al., 2006; Zeng et al., 2007]:

$$Mg^{2+} + 2\,OH^- \leftrightarrow Mg(OH)_2 \tag{26}$$

$$2\,Mg^{2+} + SiO_4^{4-} \leftrightarrow Mg_2SiO_4 \tag{27}$$

$$Mg^{2+} + SiO_3^{2-} \leftrightarrow MgSiO_3 \tag{28}$$

Surface of $Mg(OH)_2$ is always positively charged as the isoelectric point is 13.2 [McLaughlin et al., 1993]. The silicate ions are adsorbed onto $Mg(OH)_2$ via electrostatic interaction. It has been indicated that precipitation of $Mg(OH)_2$ assists in the removal of silica from RO brackish water with high magnesium, mainly through adsorption [Zeng et al., 2007]. It was confirmed in current study that $MgCl_2$ was more effective than $CaCl_2$ in removing soluble silica from construction wastewater [Tsai et al., 2011].

Polyvalent metal ions, such as aluminum and iron are also used for silica precipitation. Aluminum and iron at low concentration can react with dissolved silica to form precipitate with reaction [Sugita et al., 1999; Gabelich et al., 2005]:

$$OFeOH.H_2O + Si(OH)_4 \rightarrow Fe(OH)_3.SiO_2 + 2\,H_2O \tag{29}$$

$$2\,Al^{3+} + 2\,H_4SiO_4 + H_2O \rightarrow Al_2Si_2O_5(OH)_4\ \text{(kaolinite)} + 6\,H^+ \tag{30}$$

Iron (III) will react with freshly synthetic silica wastewater under certain condition within reaction time of 2 hours with removal efficiency of 28%. Bremere et al. [2000] showed that removal efficiency of silica increased to 81% when iron (III) salt was added to the wastewater. There are several mechanisms involved in removal of polymeric silica by addition of iron (III) salt. The reaction time was approximately 2 hours and the initial concentration of silica was 700 mg/L at 25°C and neutral pH [Bremere et al., 2000]. Concentration of total silica in wastewater decreased after reaching saturated solubility of amorphous silica. Afterward, monomeric silica will react with iron (III) to form iron silicates precipitates known to have lower solubility compared with amorphous silica. Similar to iron salt, the effective formation of aluminum silicate precipitates requires presence of dissolved aluminum and silica.

Precipitation and dissolution of silica depends on temperature as well. The solubility of silica increases with increasing temperature. Gallup et al. [2003] showed both the deposition rate and nucleation rate increased linearly with increasing temperature. Temperature also affects minerals that might be formed in precipitation of silica.

Utilization of Silicate Precipitate

Application of silica precipitation in geothermal brine was conducted by using CaO as a precipitation reagent. Ueda et al. [2003] mentioned that CaO is cheaper than other chemicals. The dosage of CaO decreased with recirculation of silica seed in treatment process. Gallup et al. [2003] used pressure reactor in which different precipitation reagents were used, such as magnesium, aluminum and iron to remove silica from brine wastewater. Several different crystals were found, such as kerolite when using magnesium salt as precipitation reagent, zeolite when using aluminum salt and smectite clay when using a combination of magnesium and iron salts. This silica precipitates produced can be reused as raw materials in industry.

Other materials have been developed to remove silica from wastewater. Zeng et al. [2007] used $MgCl_2$ and $ZnSO_4.7H_2O$ to remove silica from heavy oil wastewater. Addition of zinc improved the removal efficiency by complex formation of zinc and silicate. Silica can be adsorbed on sediment deposit that contains magnesium, and further decreased the concentration of silica as low as 50 mg/L in pH range 7 – 9 and 70 – 90°C.

3.5. Chemical Precipitation of Boron

Among different technologies, chemical precipitation is well-adapted for removal of high concentration of boron from wastewater [Xu and Jiang, 2007]. Calcium hydroxide ($Ca(OH)_2$) has been used effectively for removing boron at high concentration. The formation mechanism of calcium borate ($Ca_2B_2O_5.H_2O$) is via the dehydration reaction between borate and lime on the surface of lime, and the reaction between calcium and lime can be described as [Itakura et al., 2005]:

$$2\,Ca(OH)_2 + 2\,H_3BO_3 \rightarrow Ca_2B_2O_5.H_2O\,(s) + 4\,H_2O \quad (31)$$

However, the precipitation reaction is favorable only under high temperature. Boron concentration is reduced from 700 to lower than 50 mg/L at 90°C, lime dose of 50 g/L, and process time of 2 h [Remy et al., 2005]. Recyclable calcium borate precipitate formed when boron reacts with lime at 150°C for 14 hours [Itakura et al., 2005]. Authors indicated that re-dissolution calcium borate occurs during cooling, and phosphoric acid (H_3PO_4) has to be added to prevent re-dissolution of calcium borate by the formation of a HAP film. Hydrothermal process is applicable for wastewater containing concentrated borate, fluoroborate and fluoride ions [Itakura et al., 2006; 2007a]. Recyclable boron can be continuously removed under hydrothermal conditions [Sasai et al., 2011]. pH influences the precipitation of borate as related to its speciation. Tsai and Lo [2011] found that alkaline pH is favorable for calcium borate precipitate. Over 80% boron was recovered at pH 13, temperature of 130°C, and reaction time of 30 minutes.

To overcome long reaction time and high energy for heating, microwave technique has been found to induce higher efficiency and faster reaction [Tsai and Lo, 2011]. It is interesting to note that high dose ratio of H_3PO_4 and $Ca(OH)_2$ resulted in very low (10%) boron removal due to preferential reaction between H_3PO_4 and $Ca(OH)_2$ [Itakura et al., 2005; Tsai and Lo, 2011]. The hydrothermal process using lime at 100-200°C is applicable to detoxify and to recover different inorganic compounds, such as F, P, As, and etc [Itoh et al., 2010]. Mostly, the competition of these anions with borate for calcium ion under certain condition will decrease removal efficiency of borate.

Utilization of Calcium Borate Precipitate

Since boron mineral is widely used in manufacturing industries, such as ceramic, glass, and borax, it is possible to recover boron from wastewater. Tsai and Lo [2011] suggested that

$Ca_2B_2O_5.H_2O$ precipitate may possibly be used as a raw material for manufacturing process. Itakura et al. [2005] also indicated that the precipitate of calcium borate can be used as a raw material in borax production process.

3.6. Chemical Precipitation of Other Anions

There are limited publications on precipitation of other anions that are of environmental significance, such as selenium and molybdenum. Precipitation of selenium using conventional lime was studied, but proven to be inefficient since selenium does not form insoluble hydroxides. Ferrihydrite precipitation has also been tested in large scale, however, substantial iron addition was needed and the operational cost is high [USEPA, 2001]. Precipitation using sulfide ions itself has been developed; however, the mechanism is poorly understood. Precipitation is significantly affected by pH, where complete reaction was achieved at pH 7, with sulfide to selenium ratio between 1.8 and 11. When pH was increased to 9.5, the precipitation of selenium was incomplete, and above pH 9.5, there was no precipitates formed [Geoffroy and Demopoulos, 2011].

In alkaline or weakly acidic solution molybdenum exists as molybdate ion (MoO_4^{2-}), and with increasing acidic strength, it forms $HMoO_4^-$. Thus the removal of molybdenum using precipitation as molybdate salt is feasible. It is noted that in a very acidic condition, molybdenum exists as hydrated molybdic acid ($MoO_3.2H_2O$). Previous studies on precipitation of molybdenum from solution utilized ferric ion, strong acid, and sulphide [Freedman, 1963; LeGendre and Runnells, 1975; Yatirajam et al., 1976]. For molybdenum removal, precipitation using calcium was preferred than sodium, since calcium molybdate has higher selling price. The solution must be brought up to neutral pH to ensure the presence of molybdate ion, and with addition of $CaSO_4$ or $Ca(OH)_2$, molybdate could be precipitated in 10 minutes to form $CaMoO_4$ [Swinkels et al., 2004].

In other study, pH-controlled precipitation can remove cobalt and molybdate from electrodeposition process wastewater [Huang et al., 2004]. Both elements were precipitated at pH 4 and formed cobalt molybdate oxide hydrate and sodium cobalt molybdenum oxide afterward. In contrast with previous study, Huang et al. [2004] found at neutral pH, molybdenum redissolved back to solution, thus pH control is very important.

4. Computer Model for Precipitation

Variation of physico-chemical conditions in precipitation process sometimes makes the calculation of optimum dosage difficult and experiments are needed to determine the dosage of precipitation reagent. However, many computer models have been developed to assist tackling this problem. Several software packages of computer models, such as GEOCHEM, WATEQ4F, MINTEQ, VIS. MINTEQ and PHREEQC are available to simulate the precipitation process. These models predict thermodynamic equilibrium of various compounds in the aqueous solutions and soil solutions.

Generally, precise calculation needs a complete database containing thermodynamic formation constant of all possible complex that might be formed in aqueous solution. The thermodynamic data usually include thermodynamic constants and stoichiometric coefficients for soluble complexes, solid, and redox couples [Sposito and Mattigod, 1980]. For computer model, four different databases usually are used for thermodynamic modeling, such as minteq.dat, wateq4f.dat, phreeqc.dat and minteq.v4.dat. However, different database sources may be used in some computer programs to assess the equilibrium. For example, PHREEQC and VIS.MINTEQ include several different thermodynamic databases, such as minteq.dat, wateq4f.dat, and phreeqc.dat, so that during the simulation user can choose different source of database and hence, more comprehensive than other models. Mostly, the input data for modeling of precipitation among these computer models are similar. For example, the concentration of each component ion must be known. Operating parameters, such as temperature, pH, and ionic strength are included as well.

PHREEQC model has been widely used in prediction of precipitation. This model is user friendly because it allows user to modify thermodynamic database, and it is also relatively easy to work on. Recently, the USGS developed the PHREEQC under Windows, having new data for elements, such as silicate (Si) and strontium (Sr). This model also provides the Excel spreadsheet and graphical spreadsheet. PHREEQC was employed to calculate the specification and SI with respect to HAP precipitation [Liu et al., 2009; Song et al., 2002; Karapinar et al., 2004]. Le Guern et al. [2003] used it to predict arsenate trapping by iron oxyhydroxides and carbonates at hydrothermal spring outlet. PHREEQC can be used to predict surface adsorption and precipitation of fluoride onto calcite [Turner et al., 2005]. Wang et al. [2006] and Warmadewanthi and Liu [2009b] used PHREEQC to predict the precipitation of struvite. It was employed in assessing equilibrium of phosphate and fluoride on calcite [Gunawan et. al., 2010]. This model was also used to investigate precipitation of ferric phosphate as affected by molar ratio, Fe:P, pH, ionic strength, alkalinity, and temperature [Zhang et al., 2010]. Lu and Liu [2010] used this model to analyze precipitation of phosphate and fluoride in TFT-LCD wastewater using hybrid precipitation-microfiltration method.

MINTEQ and Visual MINTEQ find wide applications since 2000. Phosphate and ammonium in equilibrium with solid were assessed using Vis. MINTEQ [Lee et al., 2003]. Effect of magnesium on precipitation of struvite was studied using Vis. MINTEQ [Ali et al., 2005]. Battistoni et al. [2006] used Vis.MINTEQ to assess the optimal pH and to minimizing the co-precipitation mechanism of fluoride and phosphate in phosphogypsum leachate. The database of Vis.MINTEQ was modified by adding the struvite database and deleting some of slow forming minerals, such as bobierrite, tri-calcium phosphate, HAP, and OCP, so that it can be used in modeling the equilibrium concentrations of struvite in liquid swine manure [Çelen et al., 2007]. Cao and Harris [2008] used this model to analyze the effect of carbonate and magnesium on calcium phosphate precipitation. It was used to study the effect of carbonate for removing of arsenic by zero-valent iron from groundwater [Mak et al., 2009]. Application of Vis.MINTEQ also includes predicting the precipitation and adsorption of P on fly ash [Xu et al., 2010]. This model is more user friendly and more compatible with Excel spreadsheet compared with PHREEQC version 2.8. Recently, this program was upgraded to a new version of Visual MINTEQ v 3.0. The application of MINTEQA2 computer model was develop in order to predict the mineral precipitation in solid phase [Miles and Ellis, 2001; Nelson et al., 2003]. Miles and Ellis [2001] developed this model to determine optimum condition for recovery of nutrient from anaerobic digested waste. MINTEQA2 version 3.11 has also been developed to predict the specification of chemical species for struvite precipitation with different thermodynamic constant that might exist in the lagoon [Nelson et al., 2003].

GEOCHEM is a computer program for calculating the equilibrium speciation of the chemical elements in a soil solution. Sposito and Matigod [1980] developed this model for several different applications, including prediction of the concentrations of compounds in a soil solution, prediction of the fate of pollutant in soil, estimation of the effect of pH, ionic strength, redox potential, water content, or some element on the solubility of a chosen chemical element in soil solution. This model was written with FORTRAN IV and under DOS application. It is noted that the calculation should be based on 25°C and 1 atm, while the condition of charge balance is not imposed during the calculation. Based on thermodynamic database, GEOCHEM can store up to six thermodynamic stability constants for soluble complexes and up to three thermodynamic formation constants for solid for each possible metal-ligand combination. It is necessary to create a new thermodynamic data deck and transfer to the disc storage when modification of thermodynamic database is needed [Sposito and Matigod, 1980].

Rao and Prasad [1997] used WATEQ4F to predict the thermodynamic equilibrium of phosphate mineral in groundwater. Results showed that fluoroapatite and HAP were dominant minerals in the groundwater. This model was used also to analyze the speciation of silica and its precipitates in solution [Gabelich et al., 2005]. However, temperature and charge imbalance are the limitations of this model. The uncertainties of calculation will increase considerably when the temperature is higher than 25°C or in case of charge imbalance. Modification of database is allowed in this system, but it may be more difficult than other models, such as PHREEQC and Vis.MINTEQ.

Ball and Nordstrom [1991] showed that WATEQ4F solves a set of nonlinear mass action and mass balance equations using the mathematical method known as the continued fraction method. The main objective of program WATEQ4F is to calculate SI. It then will predict whether the mineral precipitates or not. Because of thermodynamic and analytical errors, there will be a range of uncertainty for the SI values. This uncertainty will vary according to both the complexity of the mineral stoichiometry and input data errors. This model used FORTRAN 77 as language program. Software can be downloaded for free at http://wwwbrr.cr.usgs.gov/projects/GWC_chemtherm/software.htm. The installations of this program can be done by open the Wq4f_Installer.exe and choose the "Unzip." Run through the command, then the installation will be completed without further intervention. To facilitate preparation of input data sets, the Excel spreadsheet file is provided with a file named as TSTCASES.XLS. The step for running this model is as follows:

1) Open the TSTCASES.XLS file provided in Excel file, then start with row 2, substitute analytical values and modeling parameters in the columns indicated by the headers, one spreadsheet row per water analysis. A value must be entered in every cell, even if it is zero. With the exception of columns A, B, and C, all values must be in numeric.
2) The input data *must not contain commas and tabs* because the file executes using comma-separated values (CSV) file and the spreadsheet-compatible output file is a tab-separated file.

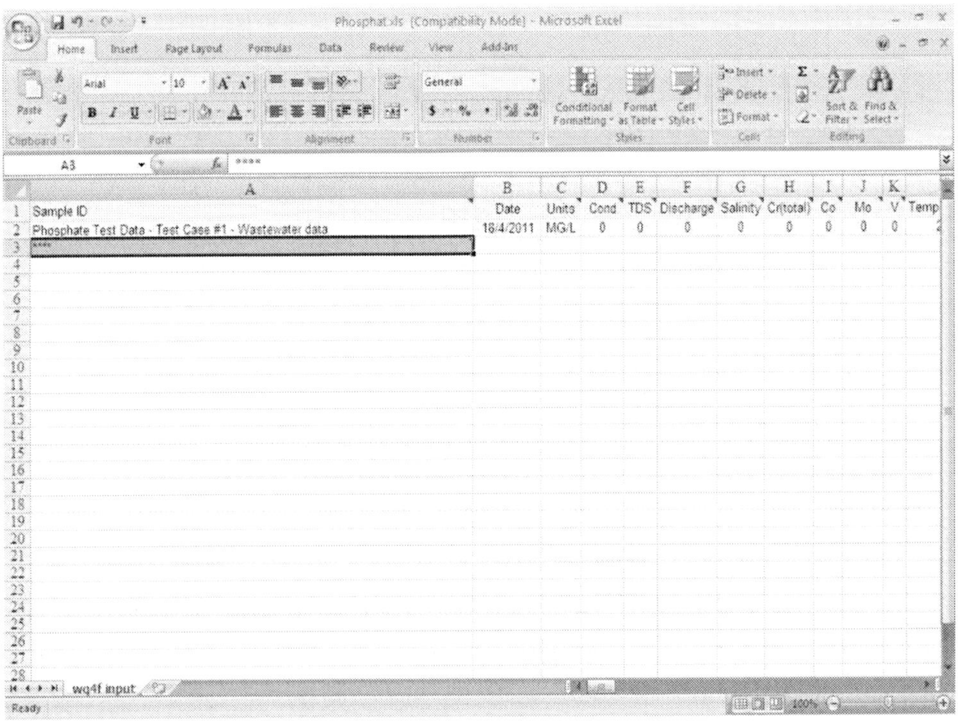

Figure 4. Spreadsheet input data for WATEQ4F.

3) The four asterisks in column 1 row 3 (Figure 4, highlighted in grey) are necessary to give END command, either to terminate single run or to write new input data for multiple runs.
4) Save the file by using Excel on CSV (MS-DOS) (filename.csv) format. The file must contain eight characters without space.
5) Continue by open the WATEQ4F program. Fill in the name of input data. It means that the *input data must be available before execution of this program.* Program will offer the name of output data and spreadsheet data.

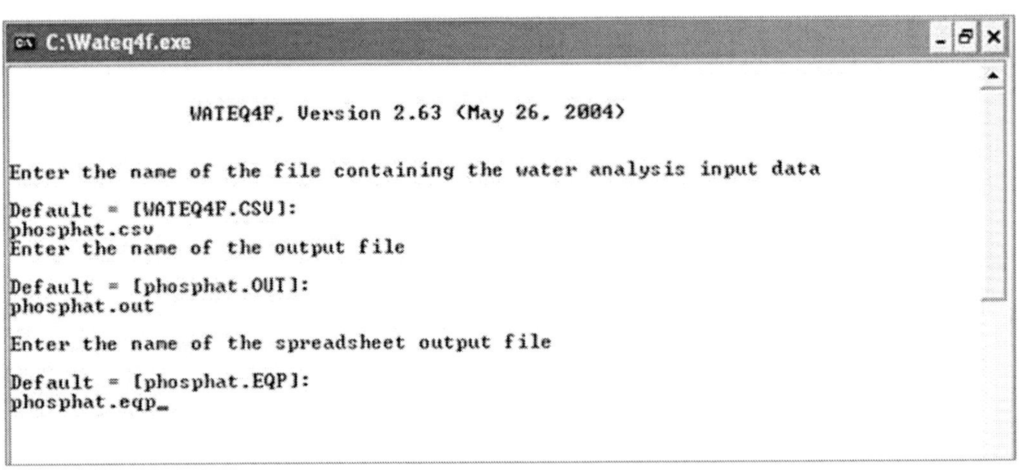

Figure 5. Input data in WATEQ4F.

6) Then the output data can be seen through the WATEQ4F.OUT by using WordPad program or by spreadsheet in Excel data through the WATEQ4F.EQP.

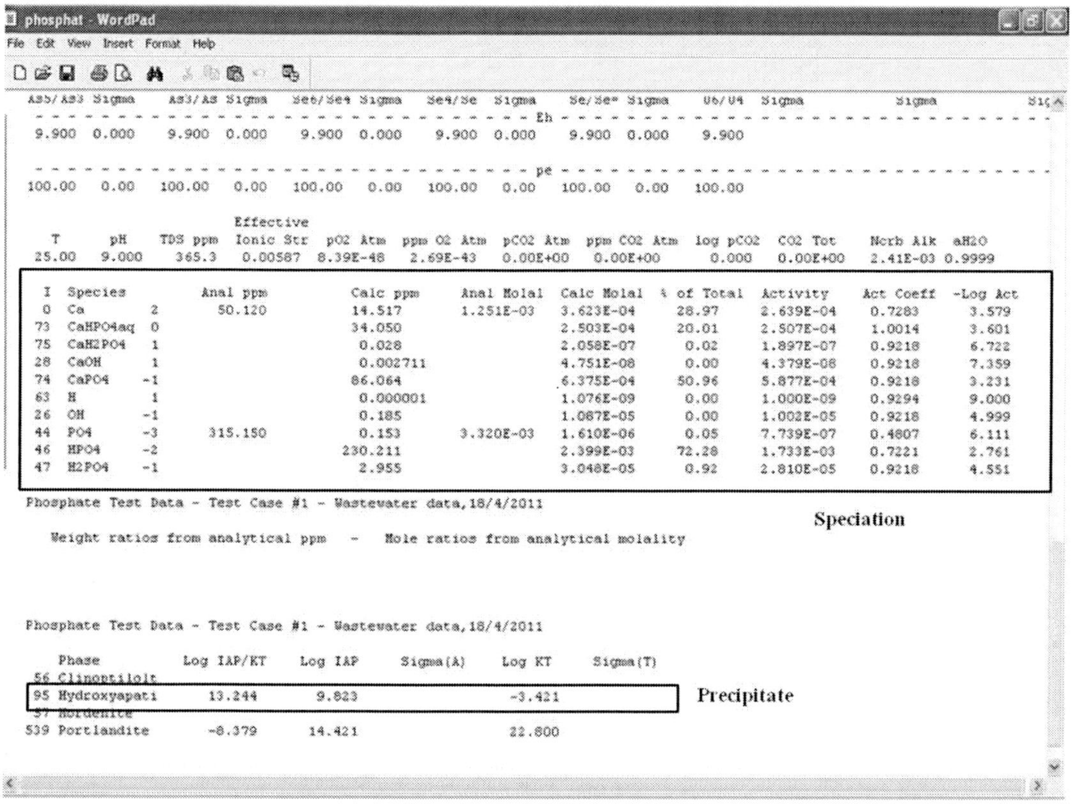

Figure 6. Output data of WATEQ4F.

Calculation results reveal that HAP forms in mineral phase, since it is shown that the value of log (IAP/KT) > 0, standing for supersaturation and formation of precipitation. The equation of SI can be seen in Equation (5). Log KT is the new solubility product of mineral in this calculation. Specification of ions can also be assessed by percentage of total species in the solution. For example, $CaHPO_4(aq)$ is found at concentration of 2.5×10^{-4} Molal or 20.01% of total concentration of ions in the solution.

MINTEQA2 can be used to calculate the equilibrium composition of dilute aqueous solutions. It can be used to calculate the mass distribution between the dissolved, adsorbed, and multiple solid phases under a variety of conditions, including gas phase with constant partial pressure [Allison et al., 1991]. This model was developed by the United States Environmental Protection Agency (USEPA) and can be downloaded freely from website: www.epa.gov/ceampubl/mmedia/minteq/. The INSTALMT.EXE file should be downloaded as a setup file for version 4.03 of the MINTEQA2 system. This file includes the MINTEQA2 software as well as the complete set of documentation. To solve the chemical equilibrium problem, MINTEQA2 uses an initial guess for the activity of each component to calculate the concentration of each species according to mass action expressions written in terms of component activities. After the equilibrium is reached, this model will calculate the SI for each possible solid with respect to the solution. This model also provides PRODEFA2, the

Problem Definition Program for MINTEQA2, an interactive program that assists the user in developing and building an input file for the MINTEQA2 model. These two programs are completely separate; however, both programs use the same thermodynamic database files. The input data of PRODEFA2 and MINTEQA2 are depicted in Figure 7. This picture shows the steps for running the MINTEQA2 model as following:

1) From prodefa2.exe, option will be shown for seed file as a template file for input data. Any DOS name can be used or just leave it blank. Second row will appear, asking for filename for MINTEQA2 input data (required); again any DOS name can be used. For example, in this demonstration PRECIPITATE1 is used as filename.
2) The program will jump to EDIT LEVEL I. MINTEQA2 has four levels of editing, each will be explained. EDIT LEVEL I displays the settings and parameters of the default problem. For precipitation model, it is very important to determine the data in level 1, such as ionic strength, concentration, the solid precipitation condition, pH, pe, and Eh. The last two notations are related to the redox reaction that showed the tendency of a solution to donate or accept a proton.

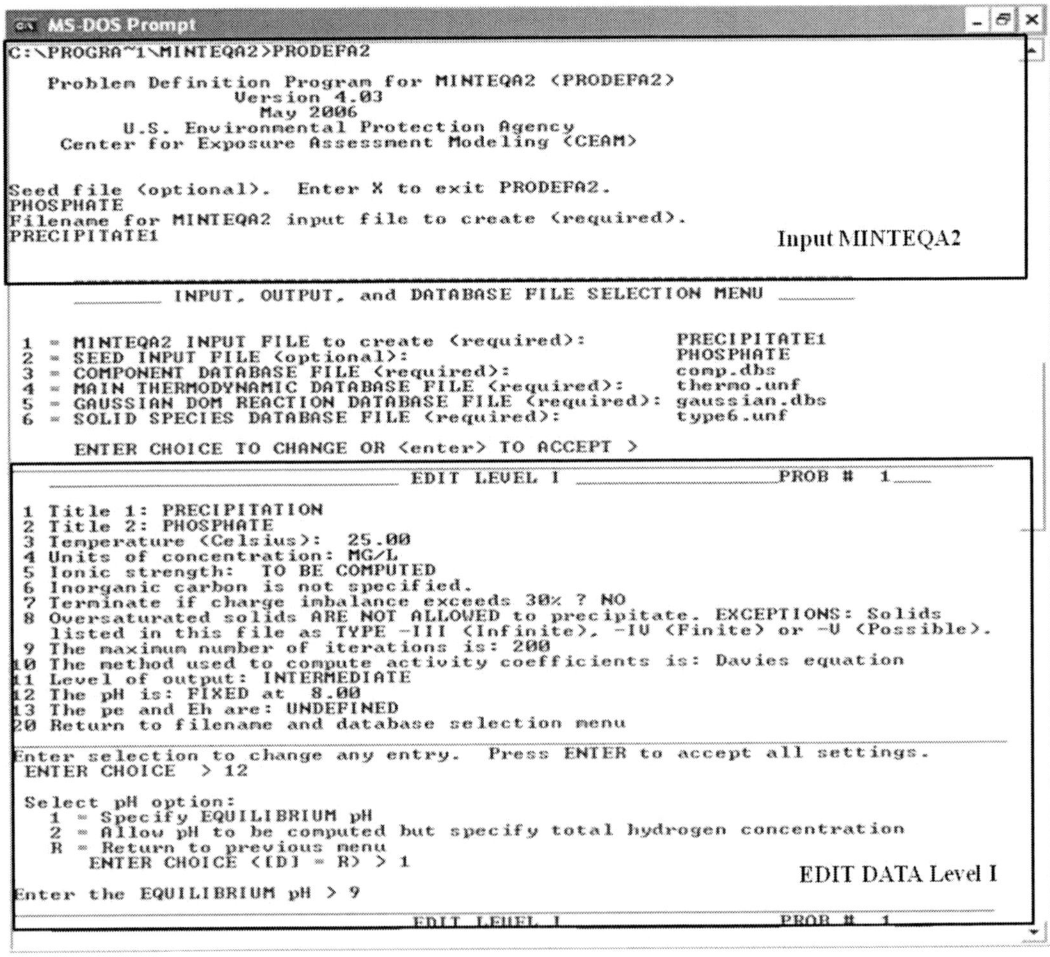

Figure 7. Input data for EDIT LEVEL I in MINTEQA2.

3) After EDIT LEVEL I is completed, EDIT LEVEL II will appear. This level is used to specify the chemistry of the system and infinite solids, gases, redox couples, finite solids, and other possible solids. It is important to determine the element to be included for precipitation in solution (**Figure 8**). For example, concentration of Ca^{2+} and PO_4^{-3} are used to determine precipitation of calcium phosphate in this case.

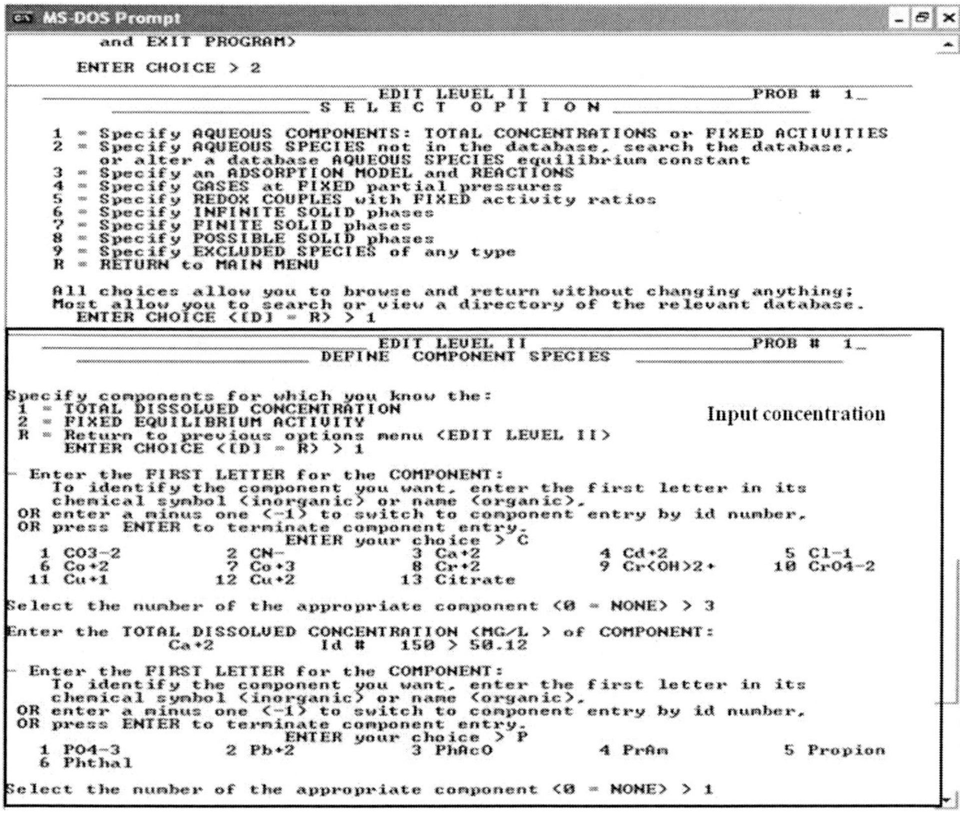

Figure 8. Input data for EDIT LEVEL II in MINTEQA2.

4) EDIT LEVEL III is used as a 'line editor', to change the name of specific component or its property. However, users are not allowed to create new component in this editing level.
5) When finished, EDIT LEVEL IV will appear. This edit level is concerned with certain utility functions rather than with chemistry. EDIT LEVEL IV is chosen if MINTEQA2 will perform a series of multiple runs as in titration or called as "sweep". The second function of this level is to direct the equilibrated mass distribution among dissolved, sorbed, and solid phase into a special file.
6) After accomplishing the model input, the model runs as in following steps:
Find minteqa2.exe file, and open it.
Then answer the filename prompts by entering the input file name previously created, and enter new name for output file.
The program will **RUN** automatically and the output data will be saved in computer folder.

7) For executing the output of this data, open *MS-DOS Prompt* and type where MINTEQA2 is saved, followed by 'edit'. For example, if MINTEQA2 is saved under Local Disk C, the command will be C:\Program Files\MINTEQA2>edit. Then a navigation box will open, and user can select the output file.

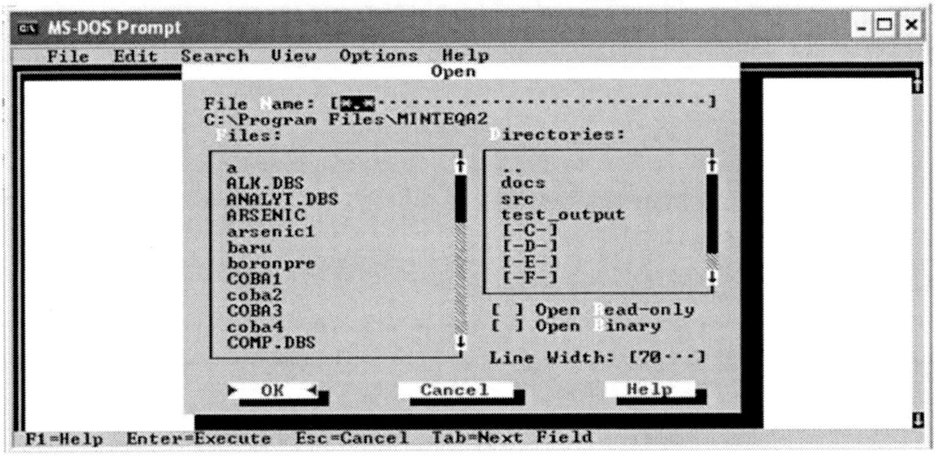

Figure 9. Navigation box for MINTEQA2 output file selection.

8) The output data will show the specification of each ion and the precipitate formed in solid phase (see **Figure 10**). For example, the output data show that HAP formed at 2.495×10^{-4} M and other undersaturated minerals are also shown in this screen. Again, the $CaHPO_4$ (aq) is 18.3 % of total phosphorus in solution.

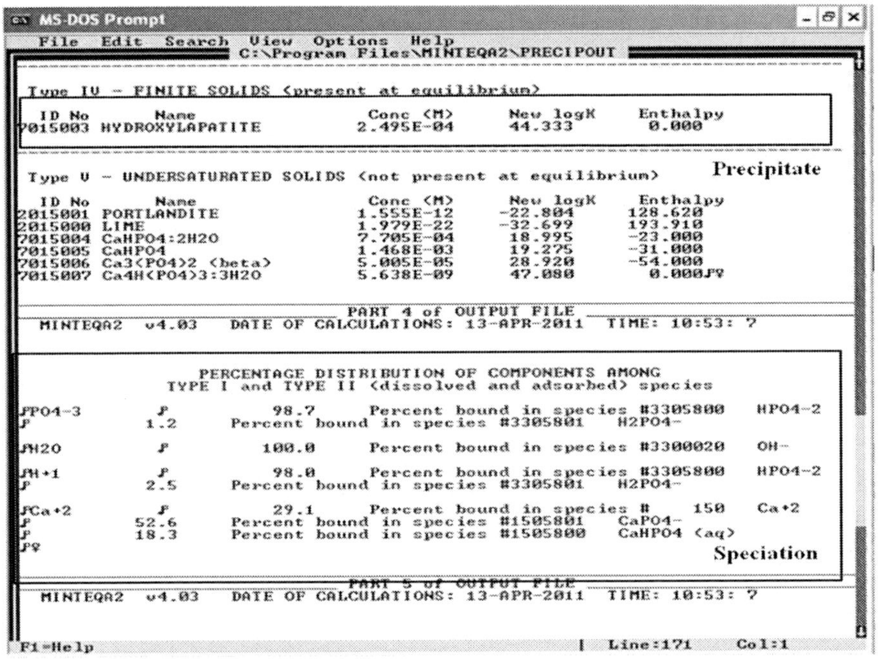

Figure 10. Example output file of MINTEQA2.

Parkhurst and Appelo [1999] developed PHREEQC model for the assessment of equilibrium chemical speciation in solution. It is based on ion association model and is capable for calculation of speciation and SI. PHREEQC calculates concentrations of elements, molalities, and activities of aqueous species, pH, pe, SI, and moles of precipitate in equilibrium. With this program, the speciation and SI of a solution with respect to different solid precipitate that might form is calculated conveniently. The main advantage of this model is that it allows users to create personal thermodynamic database, so that stoichiometric reaction and dissociation constant may be included in their database. This model can be freely downloaded from http://wwwbrr.cr.usgs.gov/projects/GWC_coupled/phreeqc/. Installation of this software is very easy through PHREEQCI application.

This model includes three different thermodynamic databases; hence users can choose their own source database for calculation. The two databases, *phreeqc.dat* and *wateq4f.dat,* are identical to the aqueous model of WATEQ4F [Ball and Nordstrom, 1991], while the other database, *minteq.dat,* is taken from MINTEQA2 [Allison et al., 1991]. However, in these databases, log Ks and enthalpies of reaction are taken from various literatures and the database provided should be considered as a preliminary reference. Users should be careful in choosing the database for their modeling.

The steps for data input are delineated as follows:

1) Initially, a blank window with a side bar will appear after opening the application. Input data for chemical system to be modeled together with the commands to perform the particular modeling task should be written in the main bar (see Figure 11).
2) For precipitation process, typical input data needed are concentration of each element (mg/L, molal, or meq/L), pH, temperature, and charge balance.
3) The mass-action equations and equilibrium constant for pure phases are defined with the PHASE data block. This PHASES block is useful for user as a command to add new thermodynamic database in modeling. For example, in this calculation the monetite, OCP, and etc were not provided in database, thus user can add corresponding stoichiometric reaction and the equilibrium constant of each reaction based on literature.
4) pH (if pH is fixed), initial composition of pure phase assemblage, and target SI are defined within the EQUILIBRIUM_PHASES data block. Equilibrium between the aqueous phase and pure phases is included in the model through heterogeneous mass-action equations. PHREEQC also allows calculation where equilibrium with a pure phase is produced by adding or removing a specified reactant (*alternative formula* and *alternative phase* command in EQUILIBRIUM_PHASES data block); the transfer of the reactant that is required to produce equilibrium with the pure phase is calculated.
5) The output of the model can be started by clicking "RUN" and then "START".
6) The "DISMISS" button appears when calculation is completed or terminated. The output folders will open by pressing it.

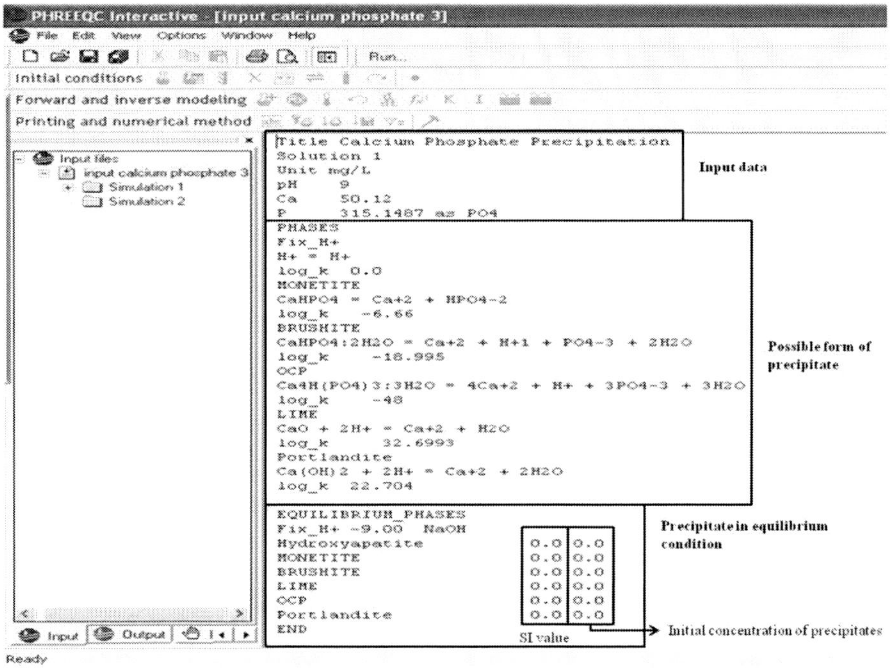

Figure 11. Input data for PHREEQC.

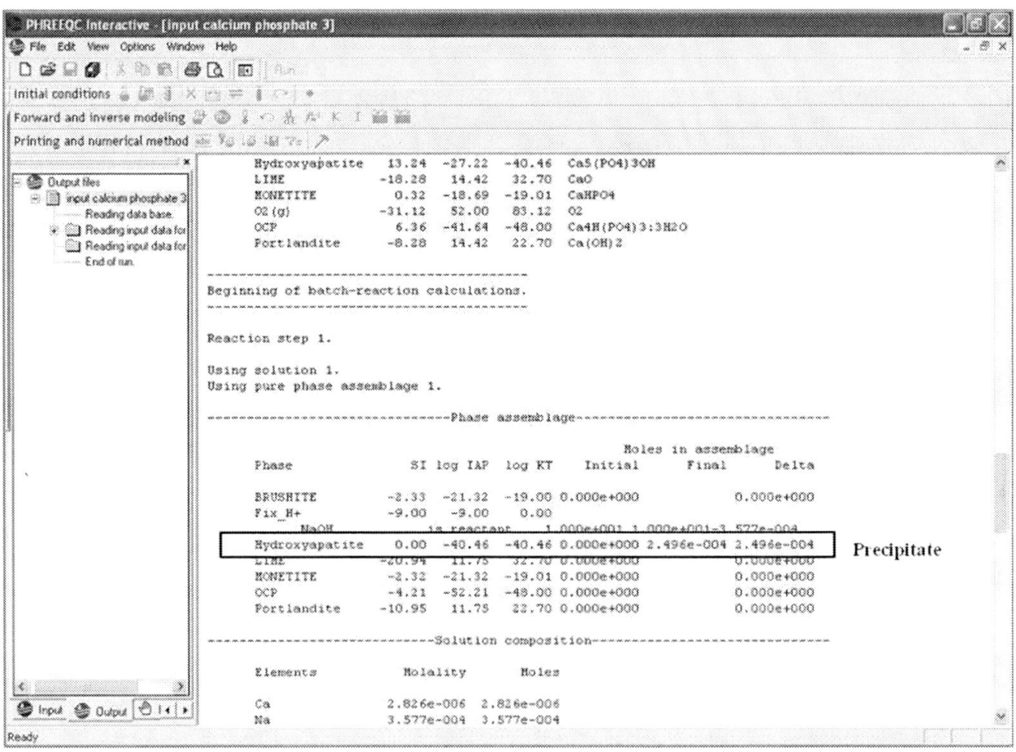

Figure 12. Output file of PHREEQC.

The output of this model will be provided in the *filename.out* file. Result shows the condition of the solution, such as ionic strength, pH, balance of ions, and etc (see Figure 12). For precipitation in solution, this calculation example shows that HAP will form in solid phase at 2.496×10^{-4} *M*. This result is similar to the results from MINTEQA2 modeling. The SI of the mineral is equal to zero when in equilibrium and it stands for saturation and precipitation will occur in solution. The amount of solid phase can be read from final moles in assemblage.

Some precipitation predictions may require some modifications when using PHREEQC. Charge balance in the solution is correlated with ionic strength and ion activity. Charge imbalance in the calculation usually is resulted from the definition of initial solution and is not changed by the subsequent equilibration with mineral. To overcome this charge imbalance, either Cl or Na may be added in solution because both ions are not strong complexer [Parkhurst and Appelo, 1999]. Special attention is given to speciation of some ions that may transform to other species which is more thermodynamically stable in solution, such as NH4+ and N2. Ammonium is involved in precipitation of struvite. Thermodynamic prediction reveals that NH4+ may form N2 because the reaction is thermodynamically favorable. However, in nature this transformation is unlikely, thus it is necessary to exclude the speciation of N2. The solubility product of N2 species is zero as included in the SOLUTION_MASTER_SPECIES data block and SOLUTION_SPECIES of PHREEQCI v 2.8 (see Figure 13). Precipitation of struvite is possible because of the availability of ammonium as shown in Figure 14.

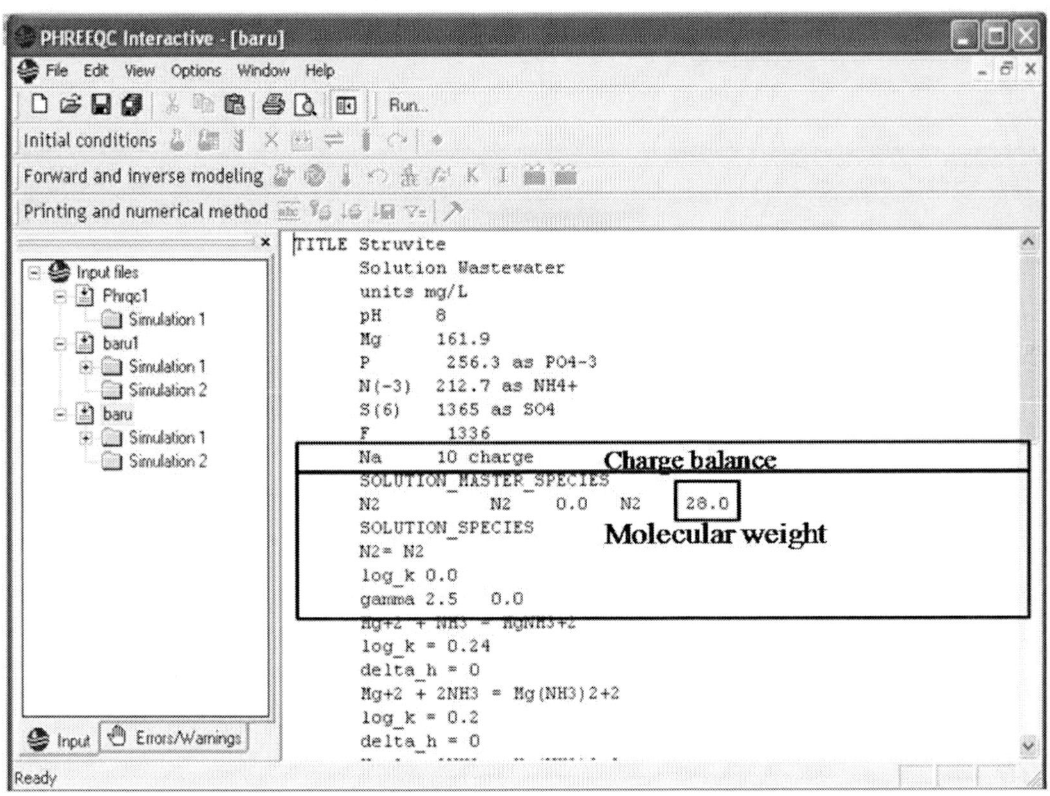

Figure 13. Modification of species in input data of PHREEQC.

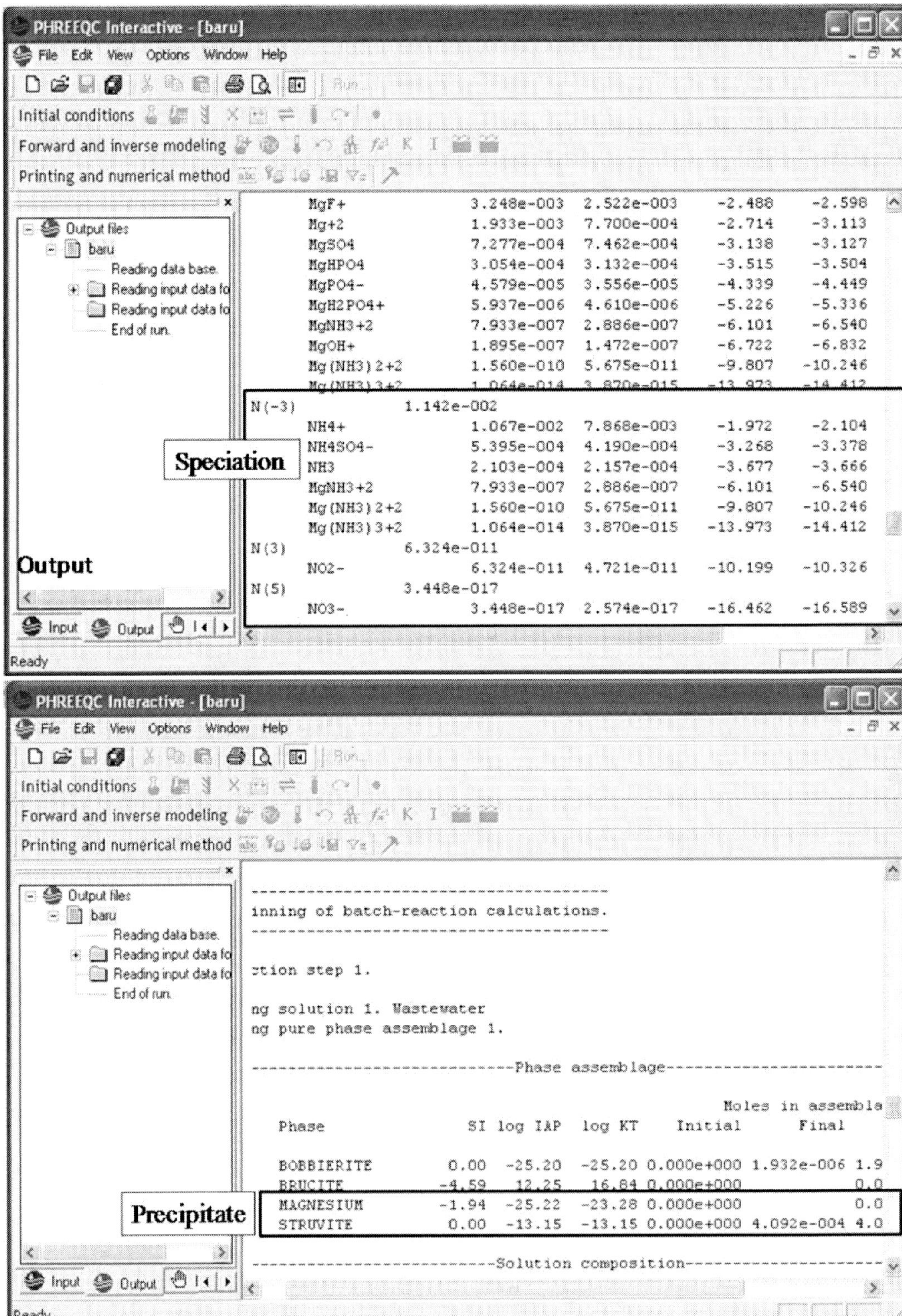

Figure 14. Output of model based on database modification (example of struvite precipitation).

Visual MINTEQ (VIS.MINTEQ) is originally the Windows version of MINTEQA2 [Çelen et al., 2007]. Gustafsson [2005] mentioned that Visual MINTEQ was developed for calculation of speciation, solubility, and equilibrium of solid and dissolved phase of minerals in aqueous solution. The advantage of this model is it allows the creation of personal database in which the possible solid phases are inserted with their characteristic values of solubility product (Ksp), enthalpy, and stoichiometric coefficients [Battistoni et al. 2006]. This model can be downloaded freely at http://www2.lwr.kth.se/English/OurSoftware/vminteq.

Typical steps for data input of VIS.MINTEQ are similar with those of PHREEQC model, except that it is under Windows version. The steps for running this program are (see Figure 15).

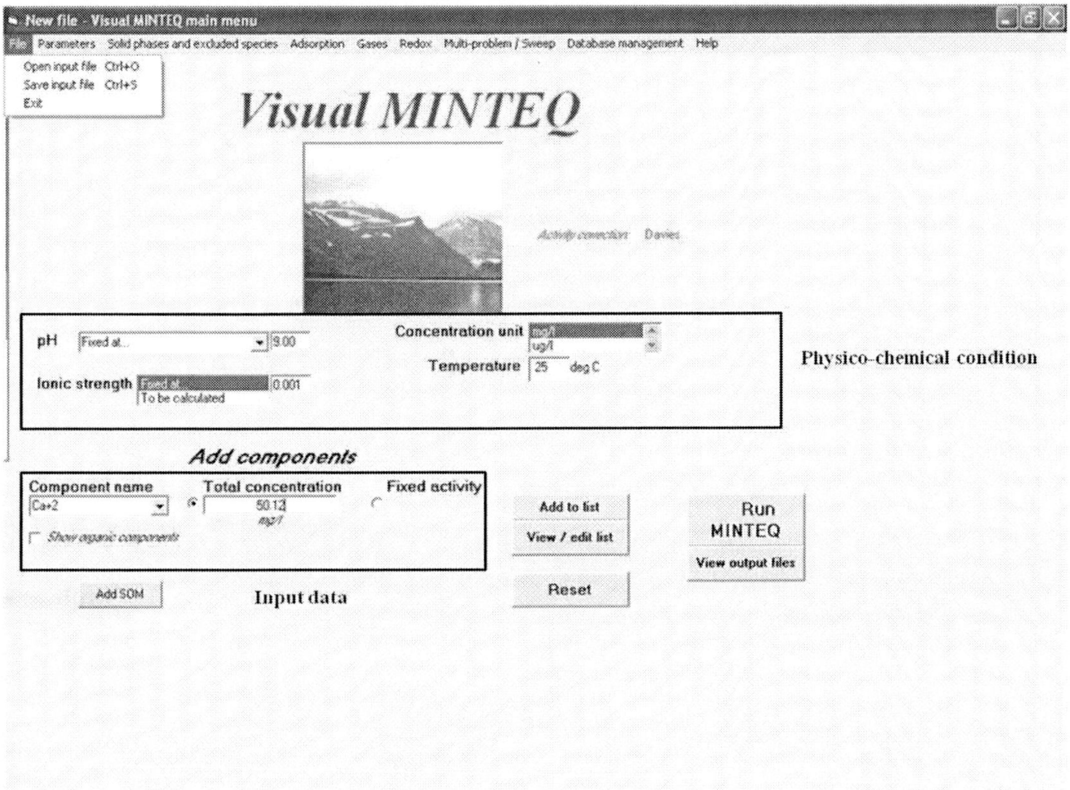

Figure 15. Input data of Visual MINTEQ.

1) Open the program VIS.MINTEQ under WINDOWS.
2) First window includes the condition of solution, such as concentration of each element (mg/L, molar, or meq/L), pH, temperature and charge balance.
3) Add component that will be included in calculation. For example, both phosphate and calcium are included to calculate calcium phosphate precipitation. After choosing the component and the concentration, continue by click "**add to list**" button.
4) If user wants to change the value of concentration of ion in the solution, it can be done through click "*add to list*" button and change the concentration.

5) "File" command is used to save input file input, and it can be used to open previous input file.
6) User can use the "Parameters" command to change the condition of calculation. For example, different equations may be used to model effect of ionic strength on precipitation of oversaturated solid, then just click "*Save and Quit*" when it is done. It is illustrated in Figure 16.

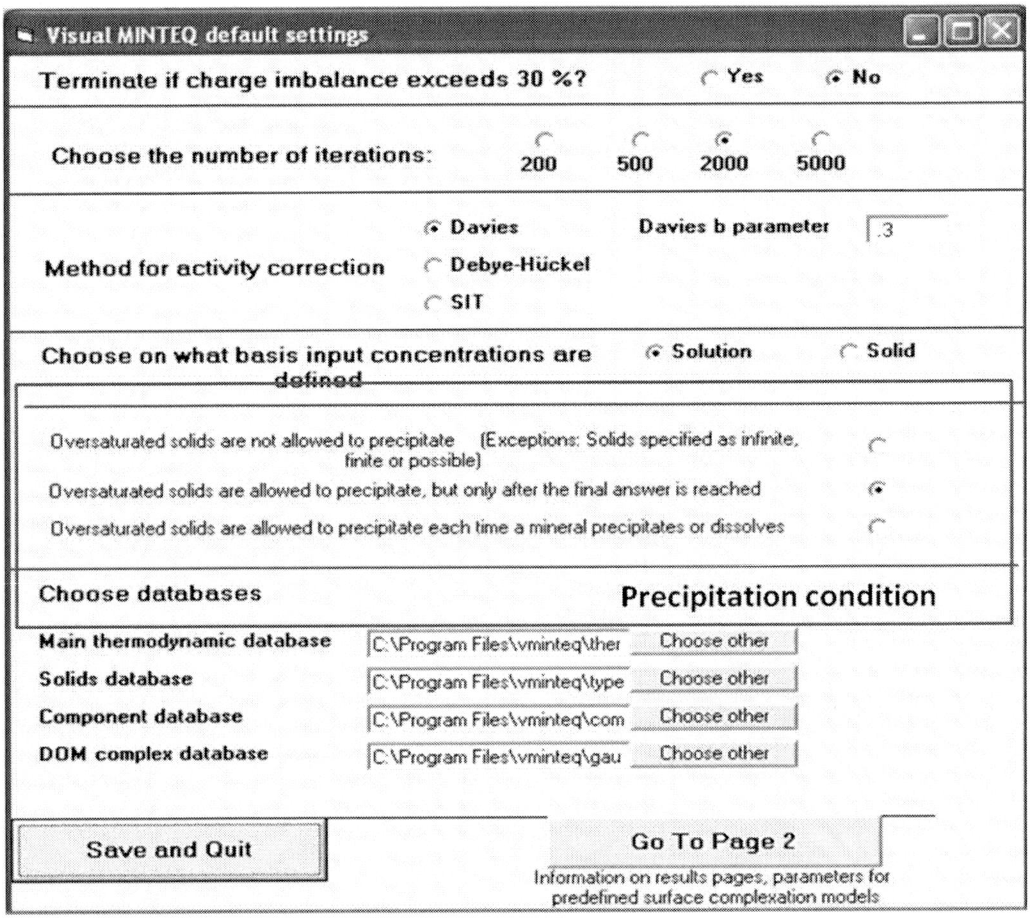

Figure 16. Edit the Visual MINTEQ solution data.

7) After get back to the main menu, one can execute the program by click "*RUN MINTEQ*", and the output data are shown by click "*View Output Data*" (see Figure 17).

Results include concentration and activity of ions in solution. The precipitate that might form in solid phase can be found by using "*AMOUNT OF FINITE SOLID*" command. This calculation, for example, shows that hydroxyapatite will form in solid phase at 2.5×10^{-4} *M*. Results are identical with those from using MINTEQA2 and PHREEQC. The fraction of chemical species can also be seen using "*EQUILIBRATED MASS DISTRIBUTION*" and

"*VIEW SPECIES DISTRIBUTION*". For example, the concentration of $CaHPO_4$ (aq) is 8.44 x 10^{-8} M and it is equivalent to 22.08% of total amount.

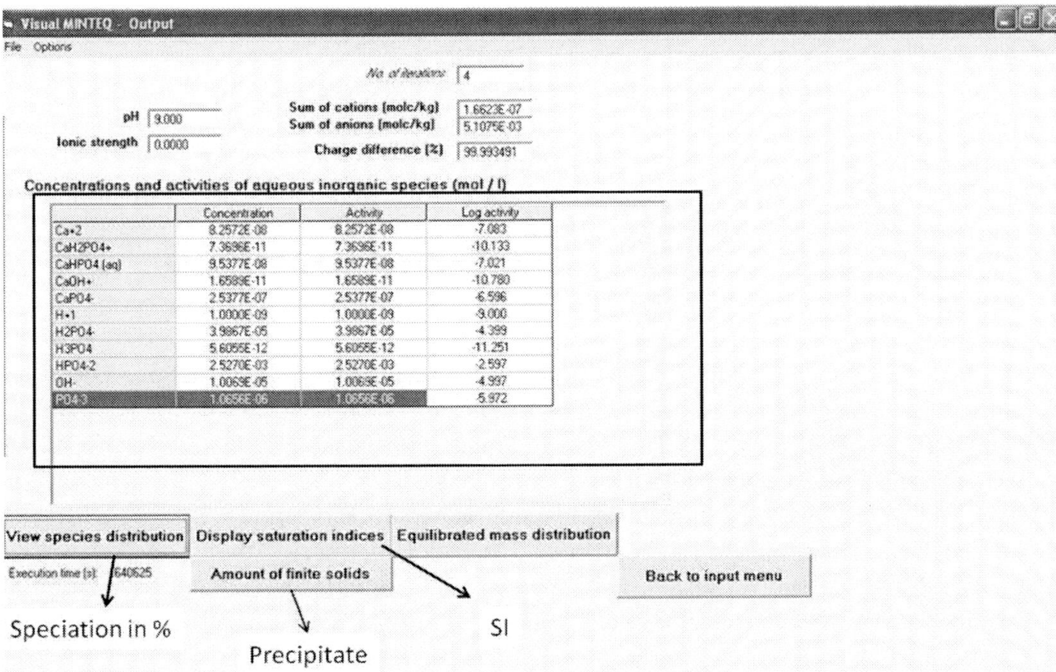

Figure 17. Output file example of VIS.MINTEQ.

CONCLUSION

The presence and chemistry of high concentration of some anions, such as arsenic, phosphate, fluoride, borate, and silicate in water and industrial wastewater were assessed. It is noted that some emerging industries, such as semiconductor, TFT-LCD, and LEDs are likely to produce more wastewater containing high concentration of anions that are of environmental significance. Since precipitation is suitable for removal of high concentration of anions from wastewater, basic theory of precipitation process and its engineering applications were examined. Several computer software commonly utilized in predicting aqueous equilibrium, particularly for precipitation, was introduced, including step-by step instruction, database available, and the comparisons.

REFERENCES

Abbona, F., H. E. Lundager Madsen, and R. Boistelle (1988), The initial phases of calcium and magnesium phosphates precipitated from solutions of high medium concentration, *Journal of Crystal Growth*, *89*(4), 592-602.

Abouzeid, A. Z. (2008), Physical and thermal treatment of phosphate ores — An overview, *International Journal of Mineral Processing*, *85*(4), 59-84.

Alam, M. G. M., E. T. Snow, and A. Tanaka (2003), Arsenic and heavy metal contamination of vegetables grown in Samta village, Bangladesh, *Science of the Total Environment*, *308*(1-3), 83-96.

Aldaco, R., A. Irabien, and P. Luis (2005), Fluidized bed reactor for fluoride removal, *Chemical Engineering Journal*, *107*(1-3), 113-117.

Aldaco, R., A. Garea, and A. Irabien (2006), Fluoride recovery in a fluidized bed: crystallization of calcium fluoride on silica sand, *Industrial and Engineering Chemistry Research*, *45*(2), 796-802.

Aldaco, R., A. Garea, and A. Irabien (2007), Calcium fluoride recovery from fluoride wastewater in a fluidized bed reactor, *Water Research*, *41*(4), 810-818.

Ali, M. A., A. B. M. Badruzzaman, M. A. Jalil, M. F. Ahmed, M. Kamruzzaman, M. A. Rahman, and A. A. Masud (2003), Fate of Arsenic in Wastes Generated From Arsenic Removal Units.

Ali, M. I., P. A. Schneider, and N. Hudson (2005), Thermodynamics and solution chemistry of struvite, *Journal of the Indian Institute of Science*, *85*(3), 141-149.

Allison, J. D., D. S. Brown, and K. J. Novo-Gradac (1991), MINTEQA2/PRODEFA2. A geochemical assessment model for environmental systems: Version 3.0. user's manual, in *Rep. EPA/600/3-91/021*, edited, U.S. Environmental Protection Agency Office of Research and Development Environmental Research Laboratory, Athens.

Andrianisa, H. A., A. Ito, A. Sasaki, J. Aizawa, and T. Umita (2008), Biotransformation of arsenic species by activated sludge and removal of bio-oxidised arsenate from wastewater by coagulation with ferric chloride, *Water Research*, *42*(19), 4809-4817.

Ann, Y., Reddy, K.R., Delfino, J.J. (2000), Influence of chemical amendments on phosphorus immobilization in soils from a constructed wetland, *Ecological Engineering*, *14*(1-2), 157-167.

Ayoob, S., and A. K. Gupta (2006), Fluoride in drinking water: A review on the status and stress effects, *Critical Reviews in Environmental Science and Technology*, *36*(6), 433-487.

Babic-Ivancic, V., J. Kontrec, L. Brecevic, and D. Kralj (2006), Kinetics of struvite to newberyite transformation in the precipitation system $MgCl_2$-$NH_4H_2PO_4$-NaOH-H_2O, *Water Research*, *40*(18), 3447-3455.

Ball, and Nordstrom (1991), User's manual for WATEQ4F, with revised thermodynamic data base and test cases for calculating speciation of major, trace, and redox elements in natural waters, edited, U.S. Geological Survey.

Banu, R. J., K. U. Do, and I. T. Yeom (2008), Phoshorus removal in low alkalinity secondary effluent using alum, *International Journal of Environment Science and Technology*, *5*(1), 93-98.

Basha, C. A., J. Selvi, E. Ramasamy, and S. Chellammal (2008), Removal of arsenic and sulphate from the copper smelting industrial effluent, *Chemical Engineering Journal*, *141*(1-3), 89-98.

Battistoni, P., G. Fava, P. Pavan, A. Musacco, and F. Cecchi (1997), Phosphate removal in anaerobic liquors by struvite crystallization without addition of chemicals: Preliminary results, *Water Research*, *31*(11), 2925-2929.

Battistoni, P., P. Pavan, M. Prisciandaro, and F. Cecchi (2000), Struvite crystallization: a feasible and reliable way to fix phosphorus in anaerobic supernatants, *Water Research*, *34*(11), 3033.

Battistoni, P., R. Boccadoro, F. Fatone, and P. Pavan (2005), Auto-nucleation and crystal growth of struvite in a demonstrative fluidized bed reactor (FBR), *Environmental Technology*, *26*(9), 975-982.

Battistoni, P., E. Carniani, V. Fratesi, P. Balboni, and P. Tornabuoni (2006), Chemical−Physical Pretreatment of Phosphogypsum Leachate, *Industrial & Engineering Chemistry Research*, *45*(9), 3237-3242.

Bhuiyan, M. I. H., D. S. Mavinic, and R. D. Beckie (2006), Solubility and thermodynamic study of struvite, *Environmental Technology*, *28*(9), 1015-1026.

Blais, J. F., Z. Djedidi, R. B. Cheikh, R. D. Tyagi, and G. Mercier (2008), Metals Precipitation from Effluents: Review, *Practice Periodical of Hazardous, Toxic, and Radioactive Waste Management*, *12*(3), 135-149.

Bothe Jr, J. V., and P. W. Brown (1999), Arsenic Immobilization by Calcium Arsenate Formation, *Environmental Science and Technology*, *33*(21), 3806-3811.

Bouropoulos, N. C., and P. G. Koutsoukos (2000), Spontaneous precipitation of struvite from aqueous solutions, *Journal of Crystal Growth*, *213*(3-4), 381.

Bremere, I., M. Kennedy, S. Mhyio, A. Jaljuli, G.-J. Witkamp, and J. Schippers (2000), Prevention of silica scale in membrane systems: removal of monomer and polymer silica, *Desalination*, *132*(1-3), 89-100.

Bristow, S., T. Shekunov, B. Y. Shekunov, and P. York (2001), Analysis of the supersaturation and precipitation process with supercritical CO2, *The Journal of Supercritical Fluids*, *21*(3), 257.

Britton, A., F. A. Koch, D. S. Mavinic, A. Adnan, W. K. Oldham, and B. Udala (2005), Pilot-scale struvite recovery from anaerobic digester supernatant at an enhanced biological phosphorus removal wastewater treatment plant, *Journal of Environmental Engineering and Science*, *4*(4), 265-277.

Burns, R. T., L. B. Moody, I. Celen, and J. R. Buchanan (2003), Optimization of phosphorus precipitation from swine manure slurries to enhance recovery, *Water Science and Technology*, *48*(1), 139-146.

Cao, X., and W. Harris (2008), Carbonate and magnesium interactive effect on calcium phosphate precipitation, *Environmental Science Technology*, *42*(2), 436-442.

Cao, X., W. G. Harris, M. S. Josan, and V. D. Nair (2007), Inhibition of calcium phosphate precipitation under environmentally-relevant conditions, *Science of The Total Environment*, *383*(1-3), 205.

Çelen, I., J. R. Buchanan, R. T. Burns, R. Bruce Robinson, and D. Raj Raman (2007), Using a chemical equilibrium model to predict amendments required to precipitate phosphorus as struvite in liquid swine manure, *Water Research*, *41*(8), 1689-1696.

Chang, M. F., and J. C. Liu (2007), Precipitation Removal of Fluoride from Semiconductor Wastewater, *Journal of Environmental Engineering*, *133*(4), 419-425.

Chen, C.-C., and Y.-C. Chung (2006), Arsenic Removal Using a Biopolymer Chitosan Sorbent, *Journal of Environmental Science and Health, Part A: Toxic/Hazardous Substances and Environmental Engineering*, *41*(4), 645-658.

Chen, S., T. Chang, and C. Lin (2006), Silica pretreatment for a RO brackish water source with high magnesium, *Water Science and Technology: Water Supply*, *6*(4), 179-187.

Chimenos, J. M., A. I. Fernández, A. Hernández, L. Haurie, F. Espiell, and C. Ayora (2006), Optimization of phosphate removal in anodizing aluminium wastewater, *Water Research*, *40*(1), 137-143.

Chimenos, J. M., A. I. Fernández, G. Villalba, M. Segarra, A. Urruticoechea, B. Artaza, and F. Espiell (2003), Removal of ammonium and phosphates from wastewater resulting from the process of cochineal extraction using MgO-containing by-product, *Water Research*, *37*(7), 1601-1607.

Chong, M. F., K. P. Lee, H. J. Chieng, and I. I. Syazwani Binti Ramli (2009), Removal of boron from ceramic industry wastewater by adsorption-flocculation mechanism using palm oil mill boiler (POMB) bottom ash and polymer, *Water Research*, *43*(13), 3326-3334.

Choong, T. S. Y., T. G. Chuah, Y. Robiah, F. L. G. Koay, and I. Azni (2007), Arsenic toxicity, health hazards and removal techniques from water: an overview, *Desalination*, *217*(1-3), 139–166

Chuang, T.-C., C. J. Huang, and J. C. Liu (2002), Treatment of Semiconductor Wastewater by Dissolved Air Flotation, *Journal of Environmental Engineering*, *128*(10), 974-980.

de-Bashan, L. E., and Y. Bashan (2004), Recent advances in removing phosphorus from wastewater and its future use as fertilizer (1997-2003), *Water Research*, *38*(19), 4222-4246.

de Luna, M. D. G., Warmadewanthi, and J. C. Liu (2009), Combined treatment of polishing wastewater and fluoride-containing wastewater from a semiconductor manufacturer, *Colloids and Surfaces A: Physicochemical and Engineering Aspects*, *347*(1-3), 64-68.

De Rooij, J. F., J. C. Heughebaert, and G. H. Nancollas (1984), A ph study of calcium phosphate seeded precipitation, *Journal of Colloid and Interface Science*, *100*(2), 350-358.

Doyle, J. D., and S. A. Parsons (2002), Struvite formation, control and recovery, *Water Research*, *36*(16), 3925-3940.

Forrest, A. L., K. P. Fattah, D. S. Mavinic, and F. A. Koch (2008), Optimizing Struvite Production for Phosphate Recovery in WWTP, *Journal of Environmental Engineering*, *134*(5), 395-402.

Freedman, M. R. (1963), Precipitation of Molybdenum (VI) in Strongly Acid Solutions, *Journal of Chemical and Engineering Data*, *8*(1), 113-116.

Fujita, T., G. Dodbiba, J. Sadaki, and A. Shibayama (2006), Removal of Anionic Metal Ions from Wastewater by Hydroxide-type Adsorbents, *The Chinese Journal of Process Engineering*, *6*(3), 357-362.

Fytianos, K., E. Voudrias, and N. Raikos (1998), Modelling of phosphorus removal from aqueous and wastewater samples using ferric iron, *Environmental Pollution*, *101*(1), 123-130.

Gabelich, C. J., W. R. Chen, T. I. Yun, B. M. Coffey, and I. H. Mel" Suffet (2005), The role of dissolved aluminum in silica chemistry for membrane processes, *Desalination*, *180*(1-3), 307-319.

Gallup, D. L., F. Sugiaman, V. Capuno, and A. Manceau (2003), Laboratory investigation of silica removal from geothermal brines to control silica scaling and produce usable silicates, *Applied Geochemistry*, *18*(10), 1597-1612.

Gaterell, M. R., R. Gay, R. Wilson, R. J. Gochin, and J. N. Lester (2000), An economic and environmental evaluation of the opportunities for substituting phosphorus recovered from wastewater treatment works in existing UK fertilizer market, *Environmental Technology*, *21*, 1067-1084.

Gemici, Ü., G. Tarcan, C. Helvacl, and A. M. Somay (2008), High arsenic and boron concentrations in groundwaters related to mining activity in the Bigadiç borate deposits (Western Turkey), *Applied Geochemistry*, *23*(8), 2462-2476.

Geoffroy, N., and G. P. Demopoulos (2011), The elimination of selenium(IV) from aqueous solution by precipitation with sodium sulfide, *Journal of Hazardous Material*, *185*(1), 148-154.

Georgantas, D. A., and H. P. Grigoropoulou (2007), Orthophosphate and metaphosphate ion removal from aqueous solution using alum and aluminum hydroxide, *Journal of Colloid and Interface Science*, *315*(1), 70-79.

Government of Indonesia (2001), Pengelolaan Kualitas Air dan Pengendalian Pencemaran Air, in *Peraturan Pemerintah Republik Indonesia No. 82*, edited.

Grzmil, B., and J. Wronkowski (2006), Removal of phosphates and fluorides from industrial wastewater, *Desalination*, *189*(1-3), 261-268.

Gunawan, E. K., Warmadewanthi, Liu, J.C. (2010), Removal of phosphate and fluoride from optoelectronic wastewater by calcite, *International Journals of Environmental Technology and Management*, *12*(2-4), 308-321.

Guo, Q., Y. Wang, and W. Liu (2008), B, As, and F contamination of river water due to wastewater discharge of the Yangbajing geothermal power plant, Tibet, China, *Environmental Geology*, *56*(1), 197-205.

Gustafsson, J. P. (2005), Visual MINTEQ, ver 2.32, edited, Royal Institute of Technology Department of Land and Water Resources Engineering, Stockholm.

Hamrouni, B., and M. Dhahbi (2001), Analytical aspects of silica in saline water -- application to desalination of brackish waters, *Desalination*, *136*(1-3), 225-232.

Hanhoun, M., L. Montastruc, C. Azzaro-Pantel, B. Biscans, M. Frèche, and L. Pibouleau (2011), Temperature impact assessment on struvite solubility product: A thermodynamic modeling approach, *Chemical Engineering Journal*, *167*(1), 50-58.

Haygarth, P. M. (1994), Global importance and global cycling of selenium, in *Selenium in the Environment*, edited by W. T. Frankenberger and S. Benson, pp. 1-28, Marcel Dekker, New York.

He, S., Y. Zhang, M. Yang, W. Du, and H. Harada (2007), Repeated use of MAP decomposition residues for the removal of high ammonium concentration from landfill leachate, *Chemosphere*, *66*(11), 2233-2238.

Hering, J. G., P. Chen, J. A. Wilkie, M. Elimelech, and S. Liang (1996), Arsenic removal by ferric chloride, *American Water Works Association Journal*, *88*(4), 155-167.

Hosni, K., S. Ben Moussa, and M. Ben Amor (2007), Conditions influencing the removal of phosphate from synthetic wastewater: influence of the ionic composition, *Desalination*, *206*(1-3), 279-285.

House, W. A. (1999), The Physico-Chemical Conditions for the Precipitation of Phosphate with Calcium, *Environmental Technology*, *20*(7), 727 - 733.

Hu, C., T. C. Zhang, Y. H. Huang, M. F. Dahab, and R. Surampalli (2005), Effects of long-term wastewater application on chemical properties and phosphorus adsorption capacity in soils of a wastewater land treatment system, *Enviromental Science and Technology*, *39*(18), 7240-7245.

Hu, C. Y., S. L. Lo, W. H. Kuan, and Y. D. Lee (2005), Removal of fluoride from semiconductor wastewater by electrocoagulation-flotation, *Water Research*, *39*(5), 895-901.

Huang, C., Jill R Pan, M. Lee, and S. Yen (2007), Treatment of high-level arsenic-containing wastewater by fluidized bed crystallization process, *Journal of Chemical Technology and Biotechnology*, *82*(3), 289-294.

Huang, C. J., and J. C. Liu (1999), Precipitate flotation of fluoride-containing wastewater from a semiconductor manufacturer, *Water Research*, *33*(16), 3403-3412.

Huang, J. H., C. Kargl-Simard, M. Oliazadeh, and A. M. Alfantazi (2004), pH-Controlled precipitation of cobalt and molybdenum from industrial waste effluents of a cobalt electrodeposition process, *Hydrometallurgy*, *75*(1-4), 77-90.

Iler, R. K. (1979), *The Chemistry, Polymerization, Colloid and Surface Properties, and Biochemistry*, John Wiley & Sons, New York.

Itakura, T., R. Sasai, and H. Itoh (2005), Precipitation recovery of boron from wastewater by hydrothermal mineralization, *Water Research*, *39*(12), 2543-2548.

Itakura, T., R. Sasai, and H. Itoh (2006), A Novel Recovery Method for Treating Wastewater Containing Fluoride and Fluoroboric Acid, *Bulletin of the Chemical Society of Japan*, *79*(8), 1303-1307.

Itakura, T., R. Sasai, and H. Itoh (2007a), In Situ Solid/Liquid Separation Effect for High-Yield Recovery of Boron and Fluorine from Aqueous Media Containing Borate or Fluoroborate Ions, *Bulletin of the Chemical Society of Japan*, *80*(10), 2014-2018.

Itakura, T., R. Sasai, and H. Itoh (2007b), Arsenic recovery from water containing arsenite and arsenate ions by hydrothermal mineralization, *Journal of Hazardous Material*, *146*(1-2), 328-333.

Itoh, H., R. Sasai, and T. Itakura (2010.), Consolidation recovery of rare/hazardous elements from polluted water by the hydrothermal mineralization process, *WIT Trans. Eco. Environ*, *140*, 369-378.

Jaffer, Y., T. A. Clark, P. Pearce, and S. A. Parsons (2002), Potential phosphorus recovery by struvite formation, *Water Research*, *36*(7), 1834-1842.

Jiang, J. Q., and N. J. D. Graham (1998), Pre-polymerised inorganic coagulant and phosphorus removal by coagulation - A review, *Water SA*, *24*(3), 237-244.

Kabay, N., I. Yilmaz, S. Yamac, S. Samatya, M. Yuksel, U. Yuksel, M. Arda, M. Saglam, T. Iwanaga, and K. Hirowatari (2004), Removal and recovery of boron from geothermal wastewater by selective ion exchange resins. I. Laboratory tests, *Reactive and Functional Polymers*, *60*, 163-170.

Kabdasli, I., S. A. Parson, and O. Tunay (2006), Effect of major ions on induction time of struvite precipitation, *Croatica Chemica Acta*, *79*(2), 243-251.

Kabir, M. Z., E. V. Emelianova, V. I. Arkhipov, M. Yunus, S. O. Kasap, and G. Adriaenssens (2006), The effects of large signals on charge collection in radiation detectors: Application to amorphous selenium detectors, *Journal of Applied Physics*, *99*(12), 12405-12401-12408.

Kalderis, D., E. Tsolaki, C. Antoniou, and E. Diamadopoulos (2008), Characterization and treatment of wastewater produced during the hydro-metallurgical extraction of germanium from fly ash, *Desalination*, *230*(1-3), 162-174.

Kapolos, J., and P. G. Koutsoukos (1999), Formation of Calcium Phosphates in Aqueous Solutions in the Presence of Carbonate Ions, *Langmuir*, *15*(19), 6557-6562.

Karapinar, N., E. Hoffmann, and H. H. Hahn (2004), Magnetite seeded precipitation of phosphate, *Water Research*, *38*(13), 3059-3066.

Kentjono, L., J. C. Liu, W. C. Chang, and C. Irawan (2010), Removal of boron and iodine from optoelectronic wastewater using Mg-Al (NO3) layered double hydroxide, *Desalination*, *262*(1-3), 280-283.

Khatibikamal, V., A. Torabian, F. Janpoor, and G. Hoshyaripour (2010), Fluoride removal from industrial wastewater using electrocoagulation and its adsorption kinetics, *Journal of Hazardous Materials*, *179*(1-3), 276-280.

Kim, D., H.-D. Ryu, M.-S. Kim, J. Kim, and S.-I. Lee (2007), Enhancing struvite precipitation potential for ammonia nitrogen removal in municipal landfill leachate, *Journal of Hazardous Materials*, *146*(1-2), 81-85.

Kim, D., J. Kim, H.-D. Ryu, and S.-I. Lee (2009), Effect of mixing on spontaneous struvite precipitation from semiconductor wastewater, *Bioresource Technology*, *100*(1), 74-78.

Kim, I., S.-H. Hong, and J.-Y. Jung (2010), Establishment of Effluent Standards for Industrial Wastewaters in Korea: Current Issues and Suggestions for Future Plan, *Journal of Water and Environment Technology*, *8*(3), 151-165.

Korchef, A., H. Saidou, and M. B. Amor (2011), Phosphate recovery through struvite precipitation by CO2 removal: Effect of magnesium, phosphate and ammonium concentrations, *Journal of Hazardous Materials*, *186*(1), 602-613.

Koutsoukos, P. G., A. N. Kofina, and P. G. Klepetsanis (2003), Exploration of alternatives for phosphorus recovery from wastewater by crystallization, paper presented at Workshop on Advance in Sensoring in Industrial Crystallization, Istanbul, Turkey.

Kunli, L., X. Lirong, T. Jian'an, W. Douhu, and X. Lianhua (2004), Selenium source in the selenosis area of the Daba region, South Qinling Mountain, China, *Environmental Geology*, *45*(3), 426-432.

Le Corre, K. S., E. Valsami-Jones, P. Hobbs, and S. A. Parsons (2005), Impact of calcium on struvite crystal size, shape and purity, *Journal of Crystal Growth*, *283*(3-4), 514-522.

Le Corre, K. S., E. Valsami-Jones, P. Hobbs, B. Jefferson, and S. A. Parsons (2007), Agglomeration of struvite crystals, *Water Research*, *41*(2), 419-425.

Le Guern, C., P. Baranger, C. Crouzet, F. Bodénan, and P. Conil (2003), Arsenic trapping by iron oxyhydroxides and carbonates at hydrothermal spring outlets, *Applied Geochemistry*, *18*(9), 1313-1323.

Lee, G., C. Chen, S.-T. Yang, and W.-S. Ahn (2010), Enhanced adsorptive removal of fluoride using mesoporous alumina, *Microporous and Mesoporous Materials*, *127*(1-2), 152-156.

Lee, S. I., S. Y. Weon, C. W. Lee, and B. Koopman (2003), Removal of nitrogen and phosphate from wastewater by addition of bittern, *Chemosphere*, *51*(4), 265-271.

LeGendre, G. R., and D. D. Runnells (1975), Removal of Dissolved Molybdenum from Wastewaters by Precipitates of Ferric Iron, *Environmental Science and Technology*, *9*(8), 744-749.

Levlin, E., Hultman, B. (2003), Phosphorus recovery from phosphate rich side streams in wastewater treatments, paper presented at Polish-Swedish Seminar Joint Polish-Swedish Reports, Gdansk Poland.

Li, Y., J. Wang, Zhaokun Luan, and Z. Liang (2010), Arsenic removal from aqueous solution using ferrous based red mud sludge, *Journal of Hazardous Material*, *177*(1-3), 131-137.

Liu, J. C., Warmadewanthi, and C.-J. Chang (2009), Precipitation flotation of phosphate from water, *Colloids and Surfaces A: Physicochemical and Engineering Aspects*, *347*(1-3), 215-219.

Lu, N. C., and J. C. Liu (2010), Removal of phosphate and fluoride from wastewater by a hybrid precipitation-microfiltration process, *Separation and Purification Technology*, *74*(3), 329-335.

Mak, M. S. H., P. Rao, and I. M. C. Lo (2009), Effects of hardness and alkalinity on the removal of arsenic(V) from humic acid-deficient and humic acid-rich groundwater by zero-valent iron, *Water Research*, *43*(17), 4296-4304.

Mamais, D., Pitt, Pa., Cheng, Y.W., Loiacono, J., Jenkins, D. (1994), Determination of ferric-chloride dose to control struvite precipitation in anaerobic sludge digesters, *Water Environment Research*, *66*(7), 912-918.

Mandal, B. K., and K. T. Suzuki (2002), Arsenic round the world: a review, *Talanta*, *58*, 201-235.

Maurer, M., and M. Boller (1999), Modelling of phosphorus precipitation in wastewater treatment plants with enhanced biological phosphorus removal, *Water Science and Technology*, *39*(1), 147-163.

Miles, A., and T. G. Ellis (2001), Struvite precipitation potential for nutrient recovery from anaerobically treated wastes, *Water Science and Technology*, *43*(11), 259-266.

Japan MOE (2007), National Effluent Standard. Ministry of the Environment, Government of Japan.

Moerman, W., M. Carballa, A. Vandekerckhove, D. Derycke, and W. Verstraete (2009), Phosphate removal in agro-industry: Pilot- and full-scale operational considerations of struvite crystallization, *Water Research*, *43*(7), 1887-1892.

Mohan, D., and C. U. Pittman Jr. (2007), Arsenic removal from water/wastewater using adsorbents—A critical review, *Journal of Hazardous Material*, *142*(1-2), 1-53.

Mohapatra, D., G. R. Chaudhury, and K. H. Park (2008), Solvent extraction approach to recover boron from wastewater generated by the LCD manufacturing industry: Part 1, *Mineral Metallurgical Process*, *25*(4), 175-180.

Montastruc, L., C. Azzaro-Pantel, B. Biscans, M. Cabassud, and S. Domenech (2003), A thermochemical approach for calcium phosphate precipitation modeling in a pellet reactor, *Chemical Engineering Journal*, *94*(1), 41-50.

Moriyama, K., T. Kojima, Y. Minawa, S. Matsumoto, and K. Nakamachi (2001), Development of artificial seed crystal for crystallization of calcium phosphate, *Environmental Technology*, *22*(11), 1245-1252.

Morse, G. K., S. W. Brett, J. A. Guy, and J. N. Lester (1998), Review: Phosphorus removal and recovery technologies, *The Science of The Total Environment*, *212*(1), 69-81.

Münch, E. V., and K. Barr (2001), Controlled struvite crystallisation for removing phosphorus from anaerobic digester sidestreams, *Water Research*, *35*(1), 151-159.

Musvoto, E. V., M. C. Wentzel, and G. A. Ekama (2000), Integrated chemical-physical processes modelling--II. simulating aeration treatment of anaerobic digester supernatants, *Water Research*, *34*(6), 1868-1880.

Nelson, N. O., R. L. Mikkelsen, and D. L. Hesterberg (2003), Struvite precipitation in anaerobic swine lagoon liquid: effect of pH and Mg:P ratio and determination of rate constant, *Bioresource Technology*, *89*(3), 229-236.

Ning, R. Y. (2003), Discussion of silica speciation, fouling, control and maximum reduction, *Desalination*, *151*(1), 67-73.

Nunez, P., H. K.Hansen, S. Aguirre, and C. Maureira (2011), Electrocoagulation ofarsenicusingironnanoparticlestotreatcoppermineral processing wastewater, *Separation and Purification Technology*, *xxx*(xxx), xxx-xxx.

Ochoa-Herrera, V., Q. Banihani, G. León, C. Khatri, J. A. Field, and R. Sierra-Alvarez (2009), Toxicity of fluoride to microorganisms in biological wastewater treatment systems, *Water Research*, *43*(13), 3177-3186.

Ohlinger, K. N., T. M. Young, and E. D. Schroeder (1998), Predicting struvite formation in digestion, *Water Research*, *32*(12), 3607-3614.

Omoike, A. I., and G. W. vanLoon (1999), Removal of phosphorus and organic matter removal by alum during wastewater treatment, *Water Research*, *33*(17), 3617-3627.

Özacar, M., and I. A. Sengil (2003), Enhancing phosphate removal from wastewater by using polyelectrolytes and clay injection, *Journal of Hazardous Materials*, *100*(1-3), 131-146.

Pakzadeh, B., and J. R. Batista (2011), Surface complexation modeling of the removal of arsenic from ion-exchange waste brines with ferric chloride, *Journal of Hazardous Material*, *188*(1-3), 399-407.

Pallier, V., G. Feuillade-Cathalifaud, and B. Serpaud (2011), Influence of organic matter on arsenic removal by continuous flow electrocoagulation treatment of weakly mineralized waters, *Chemosphere*, *83*(1), 21-28.

Parks, J.L., and M. Edwards (2005), Boron in the environment, *Crit. Rev. Environ. Sci. Technol.,* 35, 81 – 114.

Park, Y. Y., T. Tran, Y. H. Lee, Y. I. Nam, G. Senanayake, and M. J. Kim (2010), Selective removal of arsenic(V) from a molybdate plant liquor by precipitation of magnesium arsenate, *Hydrometallurgy*, *104*(2), 290-297.

Parkhurst, D., and C. Appelo (1999), User's guide to PHREEQC version 2.4.2. A computer program for speciation, batch reaction, one dimensional transport and inverse geochemical calculations, edited, U.S. Geological Survey Water Resources Investigations, Reston.

Peleka, E. N., P. P. Mavros, D. Zamboulis, and K. A. Matis (2006), Removal of phosphates from water by a hybrid flotation-membrane filtration cell, *Desalination*, *198*(1-3), 198-207.

Penetra, R. G., M. A. P. Reali, E. Foresti, and J. R. Campos (1999), Post-treatment of effluents from anaerobic reactor treating domestic sewage by dissolved-air flotation, *Water Science and Technology*, *40*(8), 137-143.

Phelan, P. J., and S. V. Mattigod (1987), Kinetics of Heterogeneously Initiated Precipitation of Calcium Phosphates1, *Soil Science Society of America Journal*, *51*(2), 336-341.

Pierri, E., D. Tsamouras, and E. Dalas (2000), Ferric phosphate precipitation in aqueous media, *Journal of Crystal Growth*, *213*(1-2), 93-98.

Rahman, M. M., Y. Liu, J.-H. Kwag, and C. Ra (2011), Recovery of struvite from animal wastewater and its nutrient leaching loss in soil, *Journal of Hazardous Materials*, *186*(2-3), 2026-2030.

Rao, N. S., and P. R. Prasad (1997), Phosphate pollution in the groundwater of lower Vamsadhara river basin, India, *Environmental Geology*, *31*(1), 117-122.

Reardon, E. J., and Y. Wang (2000), A Limestone Reactor for Fluoride Removal from Wastewaters, *Environmental Science & Technology*, *34*(15), 3247-3253.

Remy, P., H. Muhr, E. Plasari, and I. Ouerdiane (2005), Removal of boron from wastewater by precipitation of a sparingly soluble salt, *Environmental Progress*, *24*(1), 105-110.

Robins, R. G., P. M. Dove, J. D. Rimstidt, D. K. Nordstrom, and G. A. Parks (1987), Solubility and stability of scorodite, FeAsO4.2H2O, *American Mineralogist*, *72*(7-8), 842-855.

Sanchez, F., C. Gervais, A. C. Garrabrants, R. Barna, and D. S. Kossona (2002), Leaching of inorganic contaminants from cement-based waste materials as a result of carbonation during intermittent wetting, *Waste Management*, *22*(2).

Sasai, R., Y. Matsumoto, and T. Itakura (2011), Continuous-flow detoxification treatment of boron-containing wastewater under hydrothermal conditions, *Journal of the Ceramic Society of Japan*, *119*(1388), 277-281.

Schipper, W. J., A. Klapwijk, B. Potjer, W. H. Rulkens, B. G. Temmink, F. D. Kiestra, and A. C. Lijmbach (2001), Phosphate recycling in the phosphorus industry, *Environmental Technology*, *22*(11), 1337-1345.

Sheikholeslami, R., and S. Zhou (2000), Performance of RO membranes in silica bearing waters, *Desalination*, *132*(1-3), 337-344.

Sheikholeslami, R., and J. Bright (2002), Silica and metals removal by pretreatment to prevent fouling of reverse osmosis membranes, *Desalination*, *143*(3), 255-267.

Sheikholeslami, R., I. S. Al-Mutaz, T. Koo, and A. Young (2001), Pretreatment and the effect of cations and anions on prevention of silica fouling, *Desalination*, *139*(1-3), 83-95.

Shen, F., X. Chen, P. Gao, and G. Chen. (2003), Electrochemical removal of fluoride ions from industrial wastewater, *Chemical Engineering Science*, *58*(3-6), 987-993.

Shimamura, K., Tanaka, T. Miura, Y., Ishikawa, H. (2003), Development of a high-efficiency phosphorus recovery method using a fluidized-bed crystallized phosphorus removal system, *Water Science and Technology*, *48*, 163-170.

Shu, L., P. Schneider, V. Jegatheesan, and J. Johnson (2006), An economic evaluation of phosphorus recovery as struvite from digester supernatant, *Bioresource Technology*, *97*(17), 2211-2216.

Solangi, I. B., S. Memon, and M. I. Bhanger (2010), An excellent fluoride sorption behavior of modified amberlite resin, *Journal of Hazardous Materials*, *176*(1-3), 186-192.

Song, S., A. Lopez-Valdivieso, D. J. Hernandez-Campos, C. Peng, M. G. Monroy-Fernandez, and I. Razo-Soto (2006), Arsenic removal from high-arsenic water by enhanced coagulation with ferric ions and coarse calcite, *Water Research*, *40*(2), 364-372.

Song, Y. H., H. H. Hahn, and E. Hoffmann (2002), Effects of solution conditions on the precipitation of phosphate for recovery: A thermodynamic evaluation, *Chemosphere*, *48*(10), 1029-1034.

Sposito, G., and S. Matigod (1980), GEOCHEM: Computer program for the calculation of chemical equilibria in soil solutions and other natural water systems, edited, University of California, Riverside.

Strang, T., and D. G. Wareham (2006), Phosphorus removal in a waste-stabilization pond containing limestone rock filters, *Journal of Environmental Engineering and Science*, *5*(6), 447-457.

Stratful, I., M. D. Scrimshaw, and J. N. Lester (2001), Conditions influencing the precipitation of magnesium ammonium phosphate, *Water Research*, *35*(17), 4191-4199.

Stratful, I., M. D. Scrimshaw, and J. N. Lester (2004), Removal of struvite to prevent problems associated with its accumulation in wastewater treatment works, *Water Environment Research*, *76*(5), 437-443.

Stumm, W., and J. J. Morgan (1996), *Aquatic Chemistry*, 3rd ed., John Wiley and Sons Inc, New York.

Sugita, H., K. Kato, A. Ueda, I. Matsunaga, Y. Sakurai, K. Yasuda, Y. Bando, and M. Nakamura (1999), Field Tests on Silica Removal from Geothermal Brines in Sumikawa and Onuma Geothermal Areas, *JOURNAL OF CHEMICAL ENGINEERING OF JAPAN*, *32*(5), 696-700.

Suzuki, K., Y. Tanaka, K. Kuroda, D. Hanajima, Y. Fukumoto, T. Yasuda, and M. Waki (2007), Removal and recovery of phosphorous from swine wastewater by demonstration crystallization reactor and struvite accumulation device, *Bioresource Technology*, *98*(8), 1573-1578.

Swash, P. M., and A. J. Monhemius (1994), Hydrothermal precipitation from aqueous solutions containing iron (III), arsenate and sulphate, paper presented at Hydrometallurgy '94, New York.

Swinkels, P. L. J., R. D. v. d. Weijden, A. N. Ajah, Y. Arifin, H. L. Loe, M. H. Manik, I. Siriski, and M. A. Reuter (2004), Conceptual process design as a prerequisite for solving environmental problems; a case study of molybdenum removal and recovery from wastewater, *Minerals Engineering*, *17*(2), 205-215.

Tai, C. Y., P. C. Chen, and T. M. Tsao (2006), Growth kinetics of CaF2 in a pH-stat fluidized-bed crystallizer, *Journal of Crystal Growth*, *290*(2), 576-584.

Taiwan-EPA (2007), Industrial Effluent Standard, edited.

Tchobanoglous, G., F. L. Burton, and H. D. Stensel (2003), *Wastewater Engineering: Treatment and Reuse*, 4th ed., Metcalf and Eddy, Inc, New York.

Thistleton, J., T. A. Berry, P. Pearce, and S. A. Parsons (2002), Mechanisms of chemical phosphorus removal II: iron (III) salts, *Process Safety and Environmental Protection*, *80*(5), 265-269.

Thompson, G., J. Swain, M. Kay, and C. F. Forster (2001), The treatment of pulp and paper mill effluent: a review, *Bioresource Technology*, *77*(3), 275-286.

Toyoda, A., and T. Taira (2000), A new method for treating fluorine wastewater to reduce sludge and running cost, *IEEE Transactions on Semiconductor Manufacturing*, *16*, 305-309.

Tsai, H.-C., and S.-L. Lo (2011), Boron removal and recovery from concentrated wastewater using a microwave hydrothermal method, *Journal of Hazardous Materials*, *186*(2-3), 1431-1437.

Tsai, Y. W., C. P. Lu, L. S.S., and J. C. Liu (2011), Treatment and reuse of tunnel construction wastewater, *Separation and Purification Technology* (in press).

Turek, M., P. Dydo, J. Trojanowska, and B. Bandura (2007), Electrodialytic treatment of boron-containing wastewater, *Desalination*, *205*(1-3), 185-191.

Turek, M., P. Dydo, J. Ciba, J. Trojanowska, J. Kluczka, and B. Palka-Kupczak (2005), Electrodialytic treatment of boron-containing wastewater with univalent permselective membranes, *Desalination*, *185*(1-3), 139-145.

Türker, M., and I. Çelen (2007), Removal of ammonia as struvite from anaerobic digester effluents and recycling of magnesium and phosphate, *Bioresource Technology*, *98*(8), 1529-1534.

Turner, B. D., P. Binning, and S. L. S. Stipp (2005), Fluoride Removal by Calcite: Evidence for Fluorite Precipitation and Surface Adsorption, *Environmental Science & Technology*, *39*(24), 9561-9568.

Udert, K. M., T. A. Larsen, and W. Gujer (2003), Estimating the precipitation potential in urine-collecting systems, *Water Research*, *37*(11), 2667-2677.

Ueda, A., K. Kato, K. Mogi, E. Mroczek, and I. A. Thain (2003), Silica removal from Mokai, New Zealand, geothermal brine by treatment with lime and a cationic precipitant, *Geothermics*, *32*(1), 47-61.

Uludag-Demirer, S., G. N. Demirer, and S. Chen (2005), Ammonia removal from anaerobically digested dairy manure by struvite precipitation, *Process Biochemistry*, *40*(12), 3667-3674.

Urbansky, E. T. (2002), Fate of fluorosilicate drinking water additives, *Chemical Reviews*, *102*(8), 2837-2854.

USEPA (2000), Chemical Precipitation, edited by USEPA, Washington.

USEPA (2001), Final Report Selenium Treatment/Removal Alternatives Demonstration Project, edited by USEPA, Washington.

van den Broeck, K., van Hoornick, N., van Hoeymissen, J.V., de Boer, R., Giesen, A., Wilms, D. (2003), Sustainable treatment of HF wastewaters from semiconductor industry with a fluidized bed reactor, *IEEE Transactions on Semiconductor Manufacturing*, *16*(3), 423-428.

van der Houwen, J. A., and E. Vaisami-Jones (2001), The application of calcium phosphate precipitation chemistry to phosphorus recovery: the influence of organic ligands, *Environmental Technology*, *22*(11), 1325-1335.

van Kemenade, M. J. J. M., and P. L. de Bruyn (1987), A kinetic study of precipitation from supersaturated calcium phosphate solutions, *Journal of Colloid and Interface Science*, *118*(2), 564-585.

Wajima, T., Y. Umeta, S. Narita, and K. Sugawara (2009), Adsorption behavior of fluoride ions using a titanium hydroxide-derived adsorbent, *Desalination*, *249*(1), 323-330.

Wang, H.-J., W.-X. Gong, R.-P. Liu, H.-J. Liu, and J.-H. Qu (2011), Treatment of high arsenic content wastewater by a combined physical–chemical process, *Colloids and Surfaces A: Physicochemical and Engineering Aspects*, *379*(1-3), 116-120.

Wang, J., J. G. Burken, X. Zhang, and R. Surampalli (2005), Engineered Struvite Precipitation: Impacts of Component-Ion Molar Ratios and pH, *Journal of Environmental Engineering*, *131*(10), 1433-1440.

Wang, J., Y. Song, P. Yuan, J. Peng, and M. Fan (2006), Modeling the crystallization of magnesium ammonium phosphate for phosphorus recovery, *Chemosphere*, *65*(7), 1182-1187.

Warmadewanthi, and J. C. Liu (2009a), Selective Precipitation of Phosphate from Semiconductor Wastewater, *Journal of Environmental Engineering*, *135*(10), 1063-1070.

Warmadewanthi, and J. C. Liu (2009b), Recovery of phosphate and ammonium as struvite from semiconductor wastewater, *Separation and Purification Technology*, *64*(3), 368-373.

Warmadewanthi, and J. C. Liu (2009c), Selective separation of phosphate and fluoride from semiconductor wastewater, *Water Science and Technology*, *59*(10), 2047-2053.

WHO (2002), Environmental health criteria : Fluorides. International Program for Chemical Safety, World Health Organization.

WHO (2003), Boron in Drinking Water, in *Guidelines for drinking-water quality*, 2nd ed., World Health Organization.

Wu, K., H. Wang, R. Liu, X. Zhao, H. Liu, and J. Qu (2011), Arsenic removal from a high-arsenic wastewater using in situ formed Fe–Mn binary oxide combined with coagulation by poly-aluminum chloride, *Journal of Hazardous Material*, *185*(2-3), 990-995.

Wyatt, C. J., C. L. Fimbres, R. R. O. Mendez, and M. Grijalva (1998), Incidence of heavy metal contamination in water supplies in Northern Mexico, *Environmental Research*, *76*(2), 114-119.

Xu, K., T. Deng, J. Liu, and W. Peng (2010), Study on the phosphate removal from aqueous solution using modified fly ash, *Fuel*, *89*(12), 3668-3674.

Xu, Y., and J.-Q. Jiang (2007), Technologies for Boron Removal, *Industrial & Engineering Chemistry Research*, *47*(1), 16-24.

Yang, M., T. Hashimoto, N. Hoshi, and H. Myoga (1999), Fluoride removal in a fixed bed packed with granular calcite, *Water Research*, *33*(16), 3395-3402.

Yang, M., Y. Zhang, B. Shao, R. Qi, and H. Myoga (2001), Precipitative Removal of Fluoride from Electronics Wastewater, *Journal of Environmental Engineering*, *127*(10), 902-907.

Yapijakis, C., and L. K. Wang (2006), Treatment of Phosphate Industry Wastes, in *Handbook of Industrial And Hazardous Wastes Treatment*, edited by L. K. Wang, Y. T. Hung, H. H. Lo and C. Yapijakis, Marcel Dekker, Inc., New York.

Yatirajam, V., U. Ahuja, and L. R. Kakkar (1976), Use of sulphide precipitation in separation and determination of molybdenum(V), *Talanta*, *23*(11-12), 819-822.

Yetilmezsoy, K., and Z. Sapci-Zengin (2009), Recovery of ammonium nitrogen from the effluent of UASB treating poultry manure wastewater by MAP precipitation as a slow release fertilizer, *Journal of Hazardous Materials*, *166*(1), 260-269. Epub 2008 Nov 2018.

Yi, W. G., and K. V. Lo (2003), Phosphate recovery from greenhouse wastewater, *Journal of Environmental Science and Health, Part B: Pesticides, Food Contaminants, and Agricultural Wastes*, *38*(4), 501-509.

Yoshino, M., M. Yao, H. Tsuno, and I. Somiya (2003), Removal and recovery of phosphate and ammonium as struvite from supernatant in anaerobic digestion, *Water Science and Technology*, *48*(1), 171-178.

Zeng, Y., C. Yang, W. Pu, and X. Zhang (2007), Removal of silica from heavy oil wastewater to be reused in a boiler by combining magnesium and zinc compounds with coagulation, *Desalination*, *216*(1-3), 147-159.

Zhang, T., L. Ding, H. Ren, Z. Guo, and J. Tan (2010), Thermodynamic modeling of ferric phosphate precipitation for phosphorus removal and recovery from wastewater, *Journal of Hazardous Materials*, *176*(1-3), 444-450.

Zotov, N., and H. Keppler (2002), Silica speciation in aqueous fluids at high pressures and high temperatures, *Chemical Geology*, *184*(1-2), 71-82.

In: Precipitation: Prediction, Formation, Environmental Impact ISBN: 978-1-62100-447-9
Editors: Henry Dohring and Jeremy Dixon

Chapter 2

INTER-DECADAL TREND TURNINGS OF CLIMATE MODES AND DYNAMIC/THERMODYNAMIC STRUCTURES ASSOCIATED WITH A SERIES OF WINTER PRECIPITATION EVENTS OVER CHINA IN JANUARY 2008

Xiaohui Shi[1], Xiangde Xu[2] and Chungu Lu[3]
[1]State Key Laboratory of Severe Weather,
Chinese Academy of Meteorological Sciences, Beijing, China
[2]State Key Laboratory of Severe Weather,
Chinese Academy of Meteorological Sciences, Beijing, China
[3]NOAA Earth System Research Laboratory and
Colorado State University, Boulder, CO, U. S.

ABSTRACT

Inter-decadal climate variability is an important reference for inter-annual climate change and is also a disturbance in long term climate change. Inter-decadal climate variability has become a hot topic for climate change studies since the 1980s. In this context, a study of the change in surface air temperature (SAT) and precipitation on an inter-decadal time scale will elucidate the evolution and abrupt changes in the climate. In this chapter, we identified the spatial/temporal characteristics of the inter-decadal trend turning (ITT) of winter/summer climate modes over Chinese mainland. The research results show that the time of the "warm-wet" mode turning for winter climate is earlier than that for summer climate. Winter climate exhibits a distinctive "warm-dry" trend in North China after the late 1970s, and a slight "warm-dry" trend in Northeast China, South China, and Southwest China after the late 1980s. But a "warm-wet" trend appears in the rest four sub-areas after the early 1980s. Recently, the summer climate trends towards a "warm-dry" or "warm-wet" mode in most regions of China, except for South China, its summer climate starts a "cold-wet" trend in 1984. Furthermore, an analysis of ITTs of SAT and precipitation in global terrene indicates that warming trends have become highly significant across most of the world in the late 20th Century, and most terrestrial grids also have recorded an ITT of precipitation in either the 1970s or 1980s. Basically, global terrestrial regions have experienced either an increasingly "warm-dry" or an increasingly "warm-wet" climate. The northern American continent has become increasingly warm and dry after the last ITT. Meanwhile, the African continent has become increasingly warm and wet, with both Europe and most of Asia having the same trend. Southern South America and the west of Australia have experienced opposite trends in climate, becoming increasingly cold and wet.

On the other hand, the extreme event under global warming is also a focus in climate research. In winter of 2008, China experienced once-in-50-years (or once-in-100-years for some regions) snow and ice storms. These storms brought huge socio-economical impacts on the Chinese people and government. In this chapter, these cases were revisited and comprehensive analyses of the storms' dynamic/thermodynamic structure were conducted. These snowstorms were also compared with U.S. east coast snowstorms. The results from this study will provide insight on how to improve forecasts for these kinds of snowstorms. The analyses demonstrated that the storms exhibited classic patterns of large-scale circulation common to these types of snowstorms. However, several physical processes were found to be unique, and thought to have played crucial roles in intensifying and prolonging China's great snowstorms of 2008. These include a subtropical high over the west Pacific, an upper-level jet stream, and temperature and moisture inversions. The combined effects of these dynamic/thermodynamic structures are responsible for the development of the storms into a major disastrous event in Chinese history.

1. Inter-Decadal Trend Turnings of Climate Modes in Chinese Mainland and Global Terrene

1.1. Introduction

As a result of global warming, climate change has had an increasing impact on the environment, water resources, industrial and agricultural activities, and people's lives, thus drawing more and more attention. Surface air temperature (SAT) and precipitation, in particular, are two key concerns. There have been numerous studies on understanding SAT and precipitation variations in different regions. For example, there have been studies on the intra- to multi-decadal variability of SAT and precipitation in the United States (Steven 2003a, b), SAT variations for the United States compared with variations around the world (Hansen et al. 2001), SAT and precipitation variability in Europe (Klein Tank and Können 2003; Bordi et al. 2004; Tolika and Maheras 2005), precipitation variability in South America (Liebmann et al. 2004; Chagnon and Bras 2005; Haylock et al. 2006), climate variations in China (Liu et al. 2004; Ren et al. 2005; Bordi et al. 2006), and drought variability in west Africa, Asia, Australia, and around the world (Nicholson 2001; Barbé et al. 2002; Barlow et al. 2002; Nicholson 2004; Zou et al. 2005; Ma and Fu 2007). "Climate Change 2007: the Physical Science Basis" reported by the Intergovernmental Panel on Climate Change (IPCC) has raised concerns over climate change. Based on the latest findings on SAT and precipitation variability, the report includes an undisputed statement on global climate warming: human activity is largely responsible for the warming. Observational data collected from 1850 to 2006 show there was a warming process between the 1940s and 1960s and another warming process in the 1980s. Statistical analysis of global mean precipitation in the period 1900–2005 indicates an insignificant long term variability trend, with a wet trend before the 1950s, a dry trend for the 1950s–1990s, and another wet trend after the 1990s. The report also presents an analysis of linear trends of annual mean SAT and total precipitation in different regions during 1901–2005 and 1979–2005, indicating somewhat different linear trends bearing regional signatures (Trenberth et al. 2007).

Analyzing the linear trends of time series is a methodology frequently employed in climate change studies, allowing better understanding of basic variability trends in a given period of time. As far as a climate series with a longer temporal scale is concerned, linear trends may not be able to depict all the fluctuations in climate change. The Fourth IPCC Assessment Report points out that not all variability in SAT and precipitation is linear. Such variability, more often than not, represents varied features at different time periods (Trenberth et al. 2007). On the other hand, more and more published studies have shown that climate has noticeable inter-decadal variability. Inter-decadal climate variability is an important reference for inter-annual climate change and is also a disturbance in long term climate change. Inter-decadal climate variability has become a hot topic for climate change studies since the 1980s. It is also an important part of CLIVAR (Climate Variability and Predictability experiment of the World Climate Research Program). In this context, a study of the change in SAT and precipitation in the past 50-odd years on an inter-decadal time scale will elucidate the evolution and abrupt changes in the climate and is, therefore, of practical and scientific importance.

1.2. METHOD

The linear trend is able to reflect the overall trend of climate change over some period, but cannot describe the wavy character of climate change in a long time. Therefore, Tomé and Miranda (2004) put forward a new climate trend turning discriminatory model, a piecewise linear fitting model (PLFIM). According to the time scale of the studied problem, a minimum time interval between trend turning points is given. For different research purposes, the discriminatory conditions of trend turning could be different, such as the changing sign of linear trends between two consecutive time segments or the change extent exceeding a particular percentage. Under the precondition of the time interval between two consecutive breakpoints being equal to or greater than the minimum interval, i.e. filtering off the climatic waves whose period is less than the minimum interval, many combinations of time segments could be obtained. By computing the linear trend for each time segment in each combination respectively, and judging whether the trend turning takes place or not based on the discriminatory condition of trend turning, the best combination of segments can be determined by using the principle of statistical analyses, yielding the test result of trend turning points on the imposed time scale and the linear trend for each time segment.

Climate abrupt changes might be categorized into three types, the abrupt change of mean value, change rate, and trend (Li et al. 2002). They indicate the sudden change from a basic climate state (represented by a mean value) to another basic state, the distinctive difference in the change rates of two consecutive climatic states without obvious difference between their mean values, and the sudden change of the completely opposite trends, respectively. For the moment, methods frequently used in climate abrupt change research include low-pass filtering, running t-test, Cramer method, and M-K method (Fu and Wang 1992). However, these methods are mainly suitable for the test of the abrupt change of mean value. Methods for the test of the abrupt change of change rate and trend are very few. The two-phase linear regression mode (Solow 1987) and the two-phase quadratic curve fitting method (Ding et al. 1995), which developed on the basis of the former, are relatively effective, but the number of turning points in both methods must be subjectively given. In comparison with these two methods, the PLFIM need not give the number of turning points in advance. Subjected to the two constraints on the minimum distance between trend turning points and the changing sign between the linear trends of two consecutive segments (the abrupt change of trend) or the obvious difference of change rates exceeding a critical value (the abrupt change of change rate), the best continuous time segments can be gained. The number and positions of trend turning points computed in such a way would be more reasonable.

1.3. Inter-Decadal Turning of Winter/Summer Climate Modes in Chinese Mainland

The inter-decadal climate change in China and its formation mechanism remain to be further explored. For example, whether the inter-decadal turning of climate modes is a distinctively regional character or only an unordered rule with isolated and discrete distribution is unclear. What is the main difference between the inter-decadal turnings of winter/summer climate modes? Deeply exploring of these problems will lead to a better understanding about the evolution and abrupt changes of climate modes in China under the background of global change. This section employs PLFIM, to analyze the inter-decadal trend turning (ITT) of winter/summer temperature and precipitation at 160 stations and in eight sub-areas in Chinese mainland from 1961 to 2000, and explore regional structures of the inter-decadal turning of winter/summer climate modes in China, as well as the features of the inter-decadal turning of sub-area climate modes, so as to gain more objective and scientific understanding of inter-decadal changes of regional climate in China in the last 40 years.

1.3.1. Data

The January, 1960–December, 2000 monthly mean temperature and precipitation data from 600 surface meteorological stations provided by the National Meteorological Information Center of China Meteorological Administration (CMA) are used to compute the regionally mean values of temperature and precipitation over China and its eight sub-areas (Figure 4) and to test the ITTs of climate. The mean temperature and total precipitation of winter (DJF) /summer (JJA) at each station are computed from 1961 to 2000, and then their time series for the whole China and the eight sub-areas in the same period are taken as the spatial means.

The January, 1960–December, 2000 monthly mean temperature and rainfall data from 160 surface meteorological stations in China provided by the National Climate Center of CMA are used to calculate the single station time series of the mean temperature and total precipitation of winter/summer from 1961 to 2000. These single station time series are analyzed respectively to get characteristics of inter-decadal turning of winter/summer climate modes for each station.

1.3.2. Regional Structures of ITT of Winter/Summer Climate Modes

In order to investigate the ITTs of winter/summer temperature and precipitation in the last 40 years, the two imposed conditions are given. They are changing sign of linear trends between two consecutive time segments and a minimum 11-year interval between trend turning points. The ITTs of winter/summer mean temperature and precipitation over whole China from 1961 to 2000 are tested respectively with PLFIM (thick straight lines in Figure 1). To verify whether the test results of PLFIM can reflect the inter-decadal change characteristics of the climatic variables or not, the 11-year running means, i.e. the inter-decadal components (IC) of the four time series (thick curves in Figure 1), are also computed and compared with the test results of PLFIM.

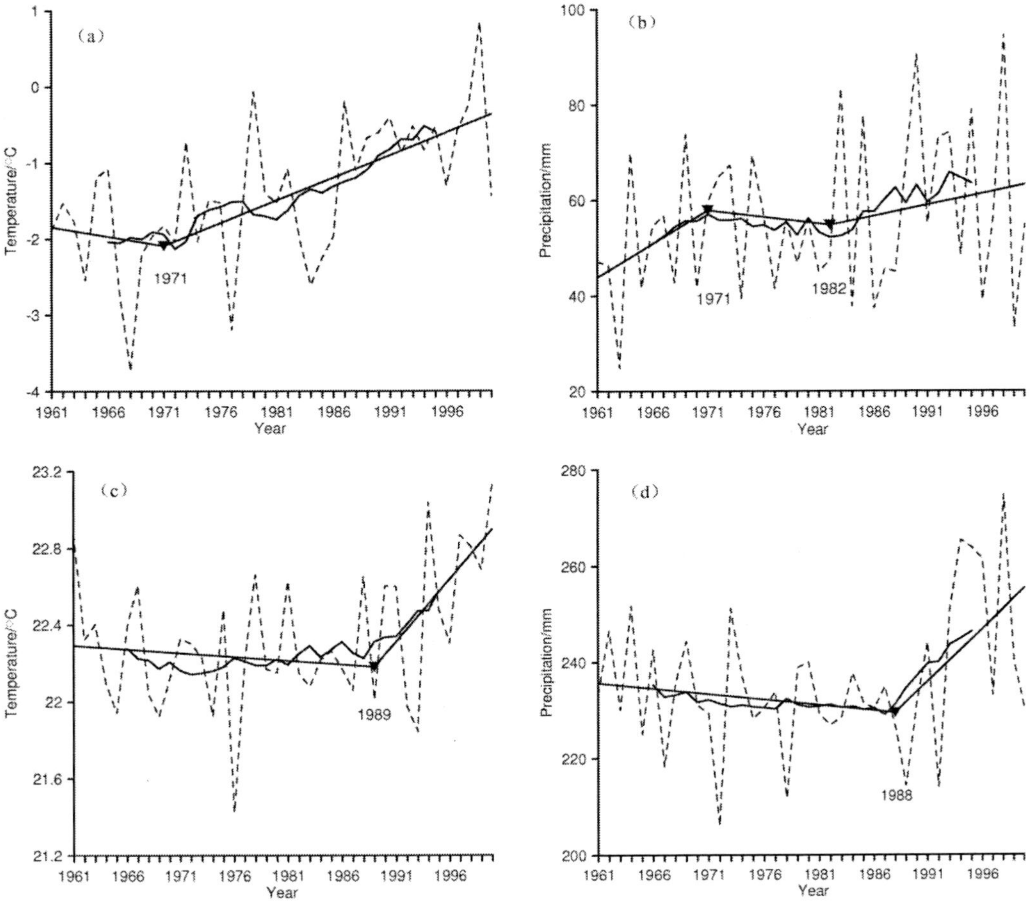

Figure 1. Time series (dash lines) of winter/summer mean temperature (a/c) and precipitation (b/d) over whole China from 1961 to 2000, their 11-year running means (thick curves), and test results of the ITT (thick straight lines). The del labeled with figures denotes the year of the ITT points.

On inter-decadal time scale, the entire-country-mean winter temperature (WT) turned from a cooling trend to a warming one in 1971, and the warming trend is more remarkable than the cooling trend. The IC of WT also confirms the robust warming trend (Figure 1a). With regard to the entire-country-mean winter precipitation (WP), two ITT points were determined in 1971 and 1982 respectively, indicating an increasing trend in the 1960s, followed by a reducing trend in the 1970s, and an increasing trend again after the early 1980s. The evolution of IC of WP is in good agreement with the test results of the ITT (Figure 1b). Therefore, it can be concluded from the above results that after the early 1980s, winter climate in China trends towards a "warm-wet" mode. The ITTs of summer temperature (ST) and precipitation (SP) are similar to each other. Both ITTs of the two climatic variables from weakly decreasing trend to remarkably increasing trend took place in the late 1980s. The ICs of ST and SP remarkably rose after the late 1980s (Figures 1c and d), that is to say, both the entire-country-mean ST and SP decreased before the late1980s, and increased afterwards. Therefore, summer climate in China trends towards a "warm-wet" mode after the late 1980s. Besides, it can be observed from Figure 1 that the trends of ICs of winter/summer temperature and precipitation are similar to the test results of PLFIM, indicating it is feasible to use the PLFIM to analyze the climatic trend, and the trend turning points mathematically computed are more objective than other methods.

The ITTs for the 1961–2000 WT and WP at 160 stations are further tested station by station with PLFIM. The occurrence time of the last ITTs and the linear trends after that (i.e. the latest linear trends) are analyzed. Figure 2a shows the results of WT at 160 stations. It can be seen that in northern China (about north of 30°N), except linearly increasing in the last 40 years without ITT at some stations, the WT at most stations had ITTs from a cooling trend to a warming trend in the early 1970s. These stations concentrate in the central and eastern parts of northern China (light gray shaded region). A few stations are mainly located in Northwest China with ITTs from a warming trend to a cooling trend in the late 1980s (black shaded regions). In southern China (about south of 30°N), the latest warming trends of WT at most stations appear after the middle 1980s (dark gray shaded region), that is to say, the ITTs with the latest warming trend occur one decade earlier at most stations in northern China than those in southern China. The situation of WP is more complicated than that of WT. The stations with an increasing trend after the last ITT mainly concentrate in the west of Northwest China, the lower reach of the Yangtze River, the east of South China, and parts of Southwest China, North China and Northeast China (dark gray shaded regions). Their last ITTs mostly took place in the early 1980s. The stations with the latest reducing trend concentrate in an SW-NE oriented zonal area, i.e. the south of Northeast China, the upper and middle reaches of the Yangtze River, the west of South China, the east of Northwest China, and the east of the Tibetan Plateau (black shaded regions). Their last ITTs mainly occurred in the late 1980s (Figure 2b).

Likewise, the ITTs for ST and SP are also tested with PLFIM, and the results are given in Figure 3. Figure 3a displays the regional characteristics of the last ITTs of ST at 160 stations in the last 40 years. It can be seen that in northern China, the last ITTs at most stations occurred in the middle and late 1980s (dark gray shaded region) and the latest linear trends are warming, except the most stations in North China, whose ITTs occurred in the early 1970s (light gray shaded region) and much earlier than that in the rest regions of northern China. In southern China, the last ITTs with a latest cooling trend happened at most stations in the 1980s (black shaded region). Figure 3b reveals the regional characteristics of the last ITTs of SP. The last ITTs with a latest increasing trend occur at most stations mainly located in South China, the lower reach of the Yangtze River, the south of Southwest China, and the west of Northwest China (dark gray regions). The last ITTs with a latest decreasing trend happen at some stations in Northeast China, central China, and parts of Northwest China and

North China (black shaded regions). Furthermore, the last ITTs of SP, with either a latest increasing trend or a latest reducing trend, mostly occurred in the 1980s.

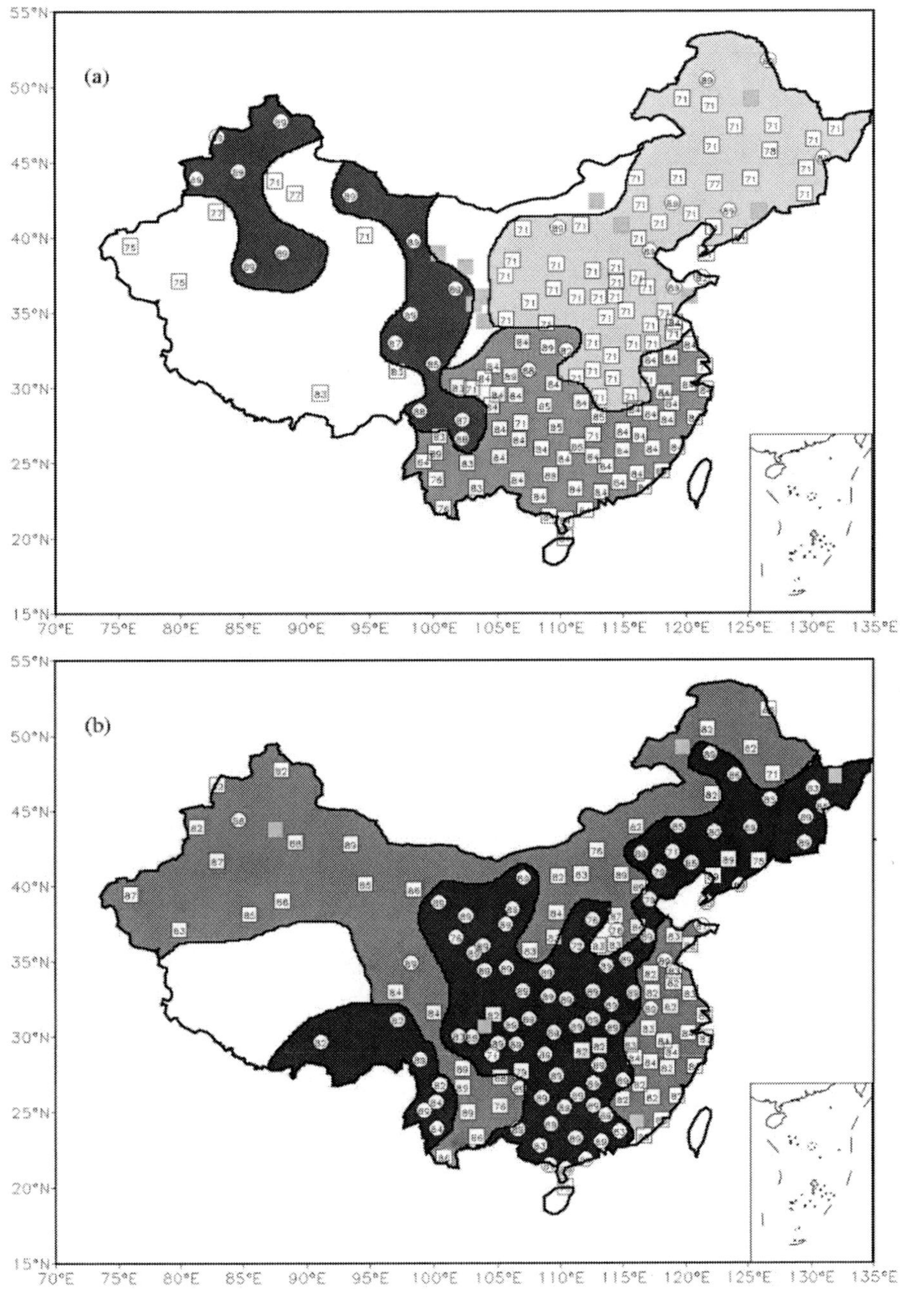

Figure 2. Last ITTs of winter temperature (a) and precipitation (b) at 160 surface meteorological stations over Chinese mainland in 1961－2000. The open circle/square denotes the decreasing/increasing trend after the last ITT respectively, and the figure within circle and square is the year of the last ITT point. The solid square denotes no ITT occurred and a linear increasing trend during 1961–2000.

The comparison between Figure 3a and Figure 3b shows that after the last ITTs, the obviously reverse trends of ST and SP appear at most stations between Northeast China and the south to the Yangtze River, that is to say, ST increases and SP reduces in Northeast China, showing a “warm-dry” trend, but in the south to the Yangtze River, ST deceases and SP increases, showing a “cooling-wet” trend. Meanwhile, the occurrence time of the last ITTs of ST and SP at most stations in these two regions is close.

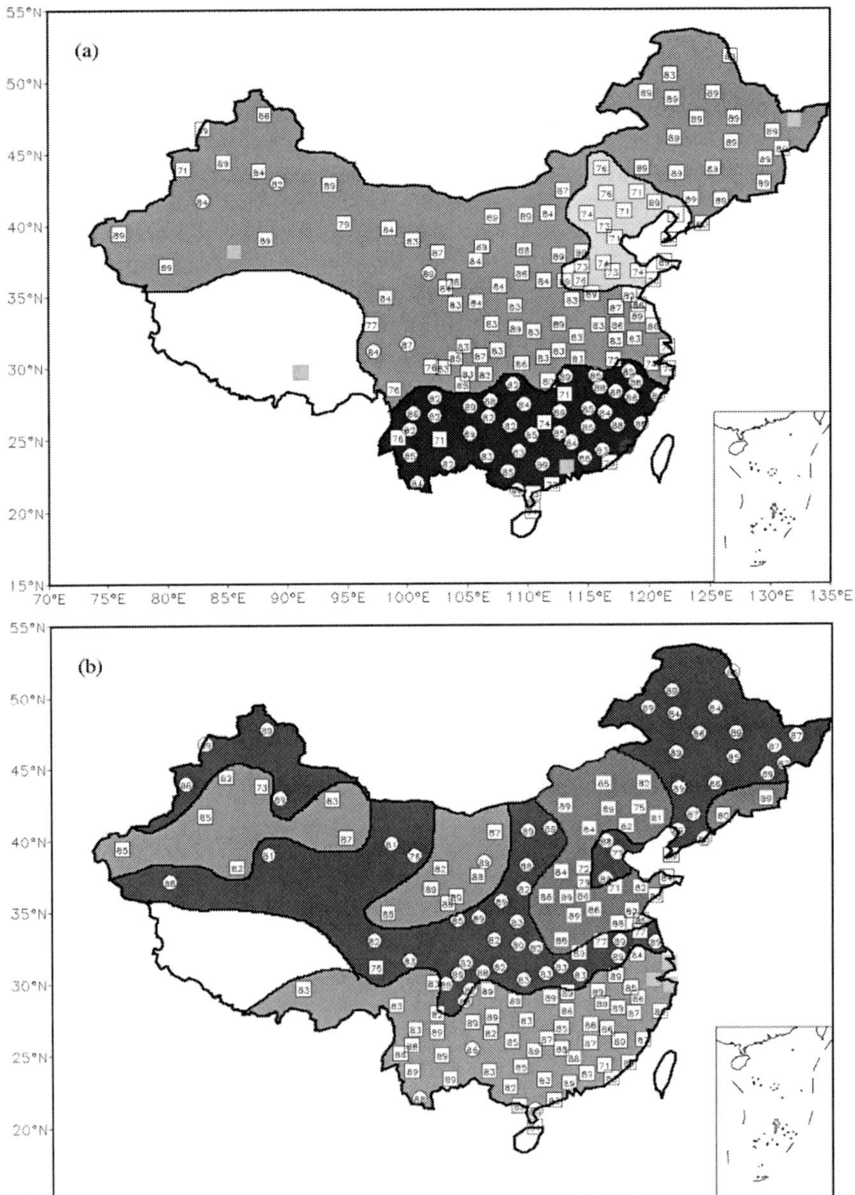

Figure 3. Same as Figure 2 except for summer temperature and precipitation. The solid circle denotes that no ITT occurred and a linear reducing trend during 1961–2000.

1.3.3. Inter-Decadal Turnings of Winter/Summer Climate Modes in Sub-Areas

From the ITT characteristics of winter/summer temperature and precipitation at 160 stations, it is easy to find that climates in various regions in China change faster than in a unanimous way, showing distinctive regional features. For the purpose of analyzing the inter-decadal turnings of sub-area winter/summer climate modes, Chinese mainland is divided into eight sub-areas primarily based on administrative divisions and the characters of the monsoon climate of China (Figure 4). They are (1) Northeast China (north of 42.5°N, east of 110°E), (2) North China (35°-42.5°N, east of 110°E), (3) the middle and lower reaches of the Yangtze River and the Huaihe River Valley, briefly Jianghuai (27.5°-35°N, east of 107.5°E), (4) South China (south of 27.5°N, east of 107.5°E), (5) Southwest China (south of 35°N, 97.5°-107.5°E), (6) the east of the Tibetan Plateau (south of 35°N, west of 97.5°E, because of the lack of stations in the west of the plateau, the analysis results for this area only represent the situations in the east of the plateau), (7) the west of Northwest China (north of 35°N, west of 97.5°E), (8) the east of Northwest China (north of 35°N, 97.5°-110°E). The division of the four sub-areas in eastern China roughly considers the northward advances of the East Asian summer monsoon as well as administrative divisions. The south of western China is divided mainly based on administrative divisions, attempting to make Southwest China cover as much more regions of Yunnan, Guizhou, Sichuan and Chongqing as possible. According to the fact that there are precipitation increasing and reducing regions respectively (Yu et al., 2003), Northwest China is divided into the east and west sub-areas. The time series of winter/summer mean temperature and total precipitation for eight sub-areas are computed using the observation data from 600 surface meteorological stations in China during 1961–2000, and their ITT characteristics are tested with the PLFIM.

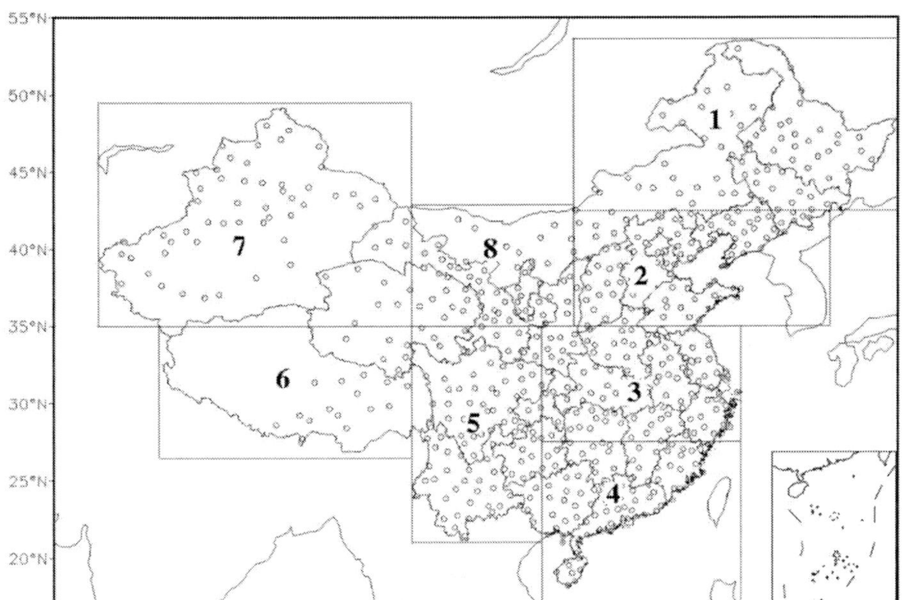

Figure 4. Positions of 600 surface meteorological stations in China and their regionalization.

Analyzing the test results of the ITTs of winter/summer temperature and precipitation in the four sub-areas in northern China, it is found that the ITT of WT in North China is consistent with that of whole China, with an ITT point from a slowly cooling trend to a rapidly warming trend in 1971 (Figure 5a). The ITT point of WP in North China from a slowly increasing trend to a slowly reducing trend was in 1979 (Figure 5b). The ITTs of WT from cooling to warming in Northeast China and the west of Northwest China also took place in the early 1970s. The WT in the east of Northwest China has no ITT, suggesting a linear increasing trend during 1961–2000. In the last 40 years, the WP in Northeast China had two ITTs. The first one from a reducing trend to an increasing trend occurred in 1975 and the second one from an increasing trend to a reducing trend happened in 1989, respectively. ITTs of the WP in the east and west of Northwest China are very consistent with each other. They showed an increasing trend before 1972, followed by a reducing trend during 1972–1983, and an increasing trend again after 1983 (Figures omitted).

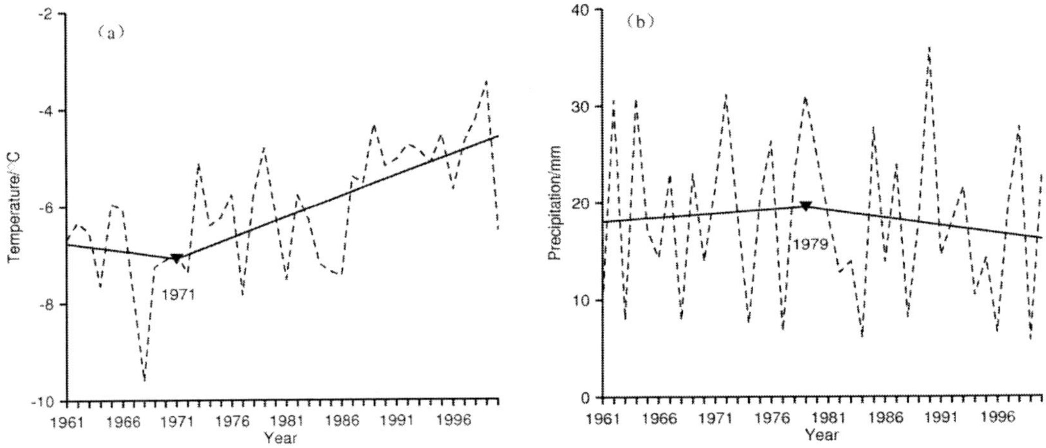

Figure 5. ITTs of winter temperature (a) and precipitation (b) in North China during 1961–2000.

With regard to the four sub-areas in southern China, the change of WT in Jianghuai is consistent with that in Northeast China, North China and whole China, with an ITT from cooling to warming in 1971. The changes of WT in Southwest China and the east of the Tibetan Plateau are unanimous, with the first ITT from warming to cooling in 1972 and the second one from cooling to warming in 1984 (Figure omitted). The change of WT in South China is similar to that in Southwest China and the east of the Tibetan Plateau, with the first ITT from warming to cooling in 1973 and the second one from cooling to warming in 1984 (Figure 6a). Comparing the test results of four sub-areas in southern China with those of four sub-areas in northern China, it can be discovered that the occurrence of the last ITTs of WT from cooling to warming in northern China is overall earlier than that in southern China, basically in agreement with the test results of 160 stations. Furthermore, the warming trends in northern China are more remarkable than those in southern China after the last ITTs.

The year 1971 is the occurrence time of the first ITT of WP from increasing to reducing in Jianghuai, and 1982 is the occurrence time of the second ITT from reducing to increasing (Figure omitted). An ITT of WP from increasing to reducing in South China occurred in 1989 (Figure 6b). The ITT of WP in Southwest China is in agreement with that in Northeast China.

It exhibits a reducing trend during the 1960s to the early 1970s, followed by an increasing trend from the early 1970s to the late 1980s, and a reducing trend again after the late 1980s. There is no ITT of WP in the east of the Tibetan Plateau in the last 40 years, showing a linear increasing trend (Figure omitted).

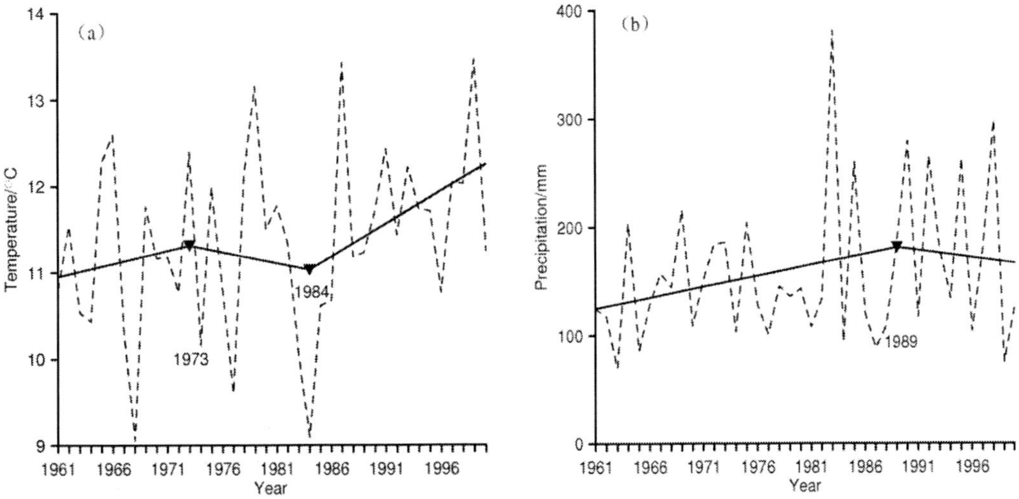

Figure 6. ITTs of winter temperature (a) and precipitation (b) in South China during 1961–2000.

According to the test results of the ITTs of WT and WP in eight sub-areas, the following conclusions can be drawn. A "warm-dry" trend of winter climate started in the late 1970s in North China and in the late 1980s in Northeast China, South China and Southwest China. A warm-wet" trend of winter climate began after the early 1980s in Northwest China, Jianghuai and the east of the Tibetan Plateau.

In the same way, the ITTs of ST and SP from 1961 to 2000 in eight sub-areas are tested with the PLFIM. It is discovered that the inter-decadal changes of ST in Northeast China and the west of Northwest China are unanimous, with 1978 being the occurrence time of the first ITT from a weak warming trend to a weak cooling trend, and 1989 being the occurrence time of the second ITT from the weak cooling trend to a distinctive warming trend. The ST in North China showed a weak cooling trend before 1989 and a distinctive warming trend after 1989. In the east of Northwest China, an ITT of ST from a cooling trend to a distinctive warming trend occurred in 1984. An ITT of ST in Jianghuai from a cooling trend to a warming trend took place in 1982. The ITTs of ST in Southwest China and the east of the Tibetan Plateau are similar. Their ITTs from a cooling trend to a warming trend occurred in 1974 and 1976, respectively (Figures omitted). Therefore, viewed from the occurrence time, the last ITTs of ST from a cooling trend to a warming trend in the three sub-areas of southern China are earlier than those in the four sub-areas of northern China, which is just opposite to the winter case. Viewed from the amplitude of temperature changes, the warming amplitudes of ST after the last ITTs in northern China are still larger than those in southern China. As a particular case, the ST in South China exhibited a cooling trend between 1961 and 1973, followed by a warming trend from 1973 to 1984, and a cooling trend again after 1984 (Figure 7a), which is basically in agreement with the research result of Wang et al. (2004). This cooling trend after 1984 might be related to the radiation forcing of aerosols brought by human activities (Chen et al. 2004).

The reducing trend of SP in North China is most prominent. Although a very weakly increasing trend appeared during 1972–1989, it is in a reducing trend as a whole. Two ITTs of SP in Northeast China took place in 1976 and 1987 respectively, with two reducing trends separated by an increasing trend, showing a wavy character. The SP in the east of Northwest China increased before 1978, and reduced afterwards (Figure omitted). The change of SP in the west of Northwest China is special. It displayed an increasing trend before the early 1970s, followed by a reducing trend during 1971－1982, and an increasing trend again after the early 1980s (Figure 7b), showing an "out-of-phase" evolutional character with respect to that of SP in the other three sub-areas of northern China. In combination with the analysis results of the ITTs of ST, it can be summed up that a "warm-dry" trend of summer climate occurred in Northeast China, the east of Northwest China and North China after the 1980s, while in the west of Northwest China, a "warm-wet" trend appeared after the late 1980s. It basically accords with the conclusions of Yu et al. (2003) and Zhou et al. (2003). Among the four sub-areas in southern China, the SP in Jianghuai has no ITT, showing a linear increasing trend in the last 40 years, that is to say, its summer climate undergoes an evolving process of "warm-dry"–"cold-wet"–"warm-wet". The SP in South China evolves in an out-of-phase with respect to its ST, with 1973 being the occurrence time of an ITT from an increasing trend to a reducing trend and 1984 being the occurrence time of the subsequent ITT from the reducing trend to an increasing trend. Consequently, its summer climate undergoes an inter-decadal evolving process of "warm-dry" –"cold-wet"–"warm-dry"–"cold-wet". The ITTs of SP in Southwest China and the east of the Tibetan Plateau are accordant with a reducing trend before 1972 and an increasing trend after 1972 (Figure omitted).

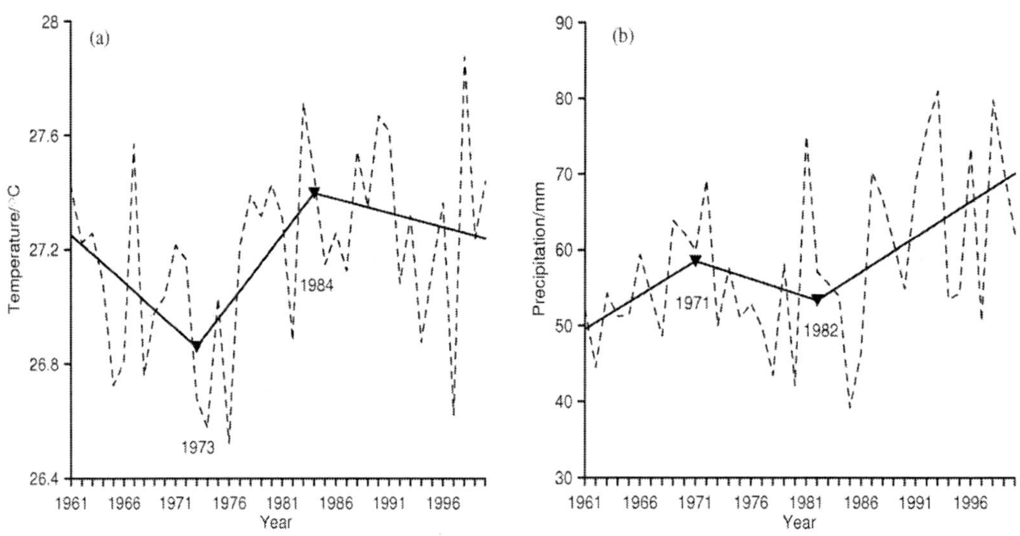

Figure 7. ITTs of summer temperatures in South China (a) and summer precipitation in the west of Northwest China (b) during 1961－2000.

The test results of the ITTs of ST and SP in sub-areas suggest that summer climate trends towards a "warm-dry" mode in Northeast China, North China and the east of Northwest

China after the late 1980s. A "warm-wet" trend occurs in Southwest China and the east of the Tibetan Plateau after the middle 1970s, and also in Jianghuai and the west of Northwest China after the early 1980s. South China is an exception, because of its summer climate trending towards a "cold-wet" mode in 1984.

1.3.4. Conclusion

Using PLFIM, comprehensive analyses are made on the regional structure of the inter-decadal turnings of climate modes at stations and the characteristics of the inter-decadal turnings of sub-area climate modes in winter/summer in Chinese mainland from 1961 to 2000. Main conclusions are as follows.

1) As far as the cases of entire-country-mean are concerned, an ITT of WT from cooling to warming occurred in 1971, while the WP undergoes two ITTs with the latest increasing trend starting from 1982. Therefore, winter climate in China trends towards a "warm-wet" mode after the early 1980s. The characteristics of the ITTs of ST and SP are similar. An ITT from weakly reducing to remarkably increasing occurred in the late 1980s, i.e. summer climate in China also trends towards a "warm-wet" mode after the late 1980s.
2) The test results of the ITTs of winter/summer temperature and precipitation at 160 surface meteorological stations in China suggest that the ITTs of WT at most stations in northern China from cooling to warming happened in the early 1970s, but in southern China in the middle 1980s. The WP at most stations in the west of Northwest China, the lower reaches of the Yangtze River, the east of South China, and parts of Southwest China, North China and Northeast China turned to an increasing trend in the early 1980s. However, the WP at most stations in the south of Northeast China, the upper and middle reaches of Yangtze River, the west of South China, the east of Northwest China, and the east of the Tibetan Plateau turned to a reducing trend in the late 1980s. The last ITTs of the ST with a latest warming trend occurred at the stations in northern China in the middle and late 1980s, while the last ITTs with a latest cooling trend took place at most stations in southern China in the 1980s. After the last ITTs in the 1980s, the SP increases at most stations in South China, the lower reaches of the Yangtze River, the south of Southwest China, and the west of Northwest China, but decreases mainly in Northeast China, central China, and parts of Northwest China and North China. The stations with close occurrence years of the last ITTs and coincident latest trends exhibit a zonal distribution, indicating remarkably regional structures.
3) The test results of ITTs of sub-area WT and WP indicate that winter climate displays a "warm-dry" trend in North China after the late 1970s, and in Northeast China, South China, and Southwest China after the late 1980s. In the rest four sub-areas, winter climate trends towards a "warm-wet" mode after the early 1980s. We also investigate the ITTs of sub-area ST and SP. It is found that summer climate trends towards a "warm-dry" mode in Northeast China, North China, and the east of Northwest China after the late 1980s. A "warm-wet" trend of summer climate occurred in Southwest China and the east of the Tibetan Plateau after the middle 1970s, as well as in Jianghuai and the west of Northwest China in the early 1980s.

South China is a particular area, whose summer climate shows a "cold-wet" trend after 1984.

4) The regional structures of ITTs at stations and the test results of eight sub-areas suggest that the last ITTs of temperature from cooling to warming in northern China occur earlier than those in southern China in winter, but later in summer. However, the latest warming trends of WT and ST in northern China are both more remarkable than those in southern China. After the last ITTs, an increasing trend of SP appears in most areas of southern China, but a reducing trend in northern China except the west of Northwest China, showing a notable "south-flood and north-drought" situation in China.

1.4. Inter-Decadal Turning of Temperature and Precipitation in Global Terrene

In last section, we used PLFIM to analyze ITT of the climate in winter and summer in Chinese mainland and obtained useful results. This section is a detailed analysis of ITTs for global terrestrial SAT and precipitation in the past –50 years, using the same model. Two conditions are preset to judge the occurrence of ITT: 1) the positive/negative signs of linear trends of two continuous time segments are different and 2) the time span between the two breakpoints is not less than 15 years.

1.4.1. Data

This study used monthly precipitation and SAT data (spatial resolution of 0.5°×0.5° and covering the period 1901–2002) provided by the University of East Anglia Climatic Research Unit (CRU) (New et al. 1999, 2000, 2002; Mitchell and Jones 2005). Only the annual total precipitation and mean SAT data for 1951–2002 have been employed in the analysis in an attempt to make the final results more reliable. Meanwhile, mainly according to the definition in the Fourth IPCC Assessment Report (Trenberth et al. 2007), the entire terrestrial system (the Antarctic is not included) was divided into 22 regions (Figure 8): 1) Alaska (ALA, 170°W–103°W, 60°N–72°N), 2) east Canada, Greenland, and Iceland (CGI, 103°W–10°W, 50°N–85°N), 3) western North America (WNA, 140°W–103°W, 30°N–60°N), 4) central North America (CNA, 103°W–85°W, 30°N–50°N), 5) eastern North America (ENA, 85°W–50°W, 25°N–50°N), 6) central North America (CAM, 116°W–83°W, 10°N–30°N), 7) northern South America (NSA, 82°W–34°W, 10°S–12°N), 8) central South America (CSA, 82°W–34°W, 30°S–10°S), 9) southern South America (SSA, 76°W–40°W, 56°S–30°S), 10) Northern Europe (NEU, 10°W–40°E, 48°N–75°N), 11) Southern Europe and the Mediterranean (SEM, 10°W–40°E, 30°N–48°N), 12) the Sahara (SAH, 20°W–60°E, 18°N–30°N), 13) West Africa (WAF, 20°W–20°E, 12°S–18°N), 14) East Africa (EAF, 20°E–52°E, 12°S–18°N), 15) South Africa (SAF, 10°E–52°E, 35°S–12 °S), 16) Northern Asia (NAS, 40°E–180°, 50°N–70°N), 17) Central Asia (CAS, 40°E–95°E, 30°N–50°N), 18) East Asia (EAS, 95°E–145°E, 20°N–50°N), 19) South Asia (SAS, 64°E–95°E, 5°N–30°N), 20) Southeast Asia (SEA, 95°E–145°E, 11°S–20°N), 21) west Australia (WAU, 110°E–136°E, 45°S–11°S), and 22) east Australia (EAU, 136°E–155°E, 45°S–11°S). We determined annual time series of SAT and total precipitation for each region in an attempt to facilitate discussion of trend changes in these regions.

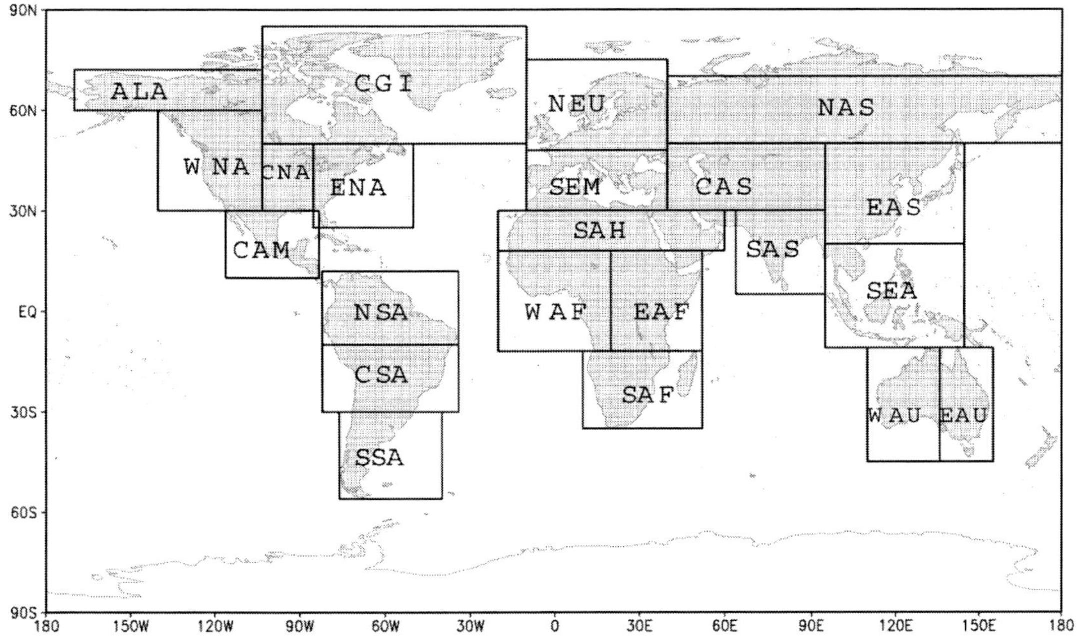

Figure 8. Distribution of CRU data and associated regions.

1.4.2. Spatial Distribution of the ITT of SAT

We examined inter-decadal trend breakpoints of annual mean SAT in 67,359 terrestrial grids for 1951–2002. We also analyzed the linear trends in and after the last ITT. Figures 9a and b show the occurrence time of ITT in annual mean SAT in each terrestrial grid, and associated linear trends after the breakpoint. It is apparent that in the past 52 years, some regions did not have an ITT in SAT, with linear warming trends seen mainly in Northern and Central Asia, West Africa, central South America, and east Australia. These regions account for 12.6% of the global terrestrial grids in the study. At the same time, a range of regions, including the central, eastern, and northwestern parts of North America, East Africa, and Northern Asia, registered an earlier ITT of SAT in the 1960s, mainly featuring a warming trend after the breakpoint. These regions account for 21.6% of the total grids in the study. In addition, 13.5% of the grids had a cooling trend after the inter-decadal breakpoint in the 1980s; the grids are mainly in western North America (42.8% of the grids in the region), southern South America (58.2% of the grids in the region), northwestern North Asia (26.0% of the grids in the region), and west Australia (57.1% of the grids in the region). The regions corresponding to the remaining grids (51.5% of the global terrestrial grids in the study) witnessed an ITT of annual mean SAT in either the 1970s or 1980s, with a warming trend after the breakpoint. In the past –50 years, the grids for which there has basically been no ITT in the SAT have had a warming trend, and 85.7% of grids in this study have had warming trends after breakpoints. It is apparent that global warming has become highly significant in the global terrestrial system in the late 20th Century. The numbers of grids with different ITT patterns (grouped according to the time of the breakpoint and trend after the breakpoint) as a proportion of the total number of grids are given in Table 1 (an underlined number indicates the largest portion of ITT in a region).

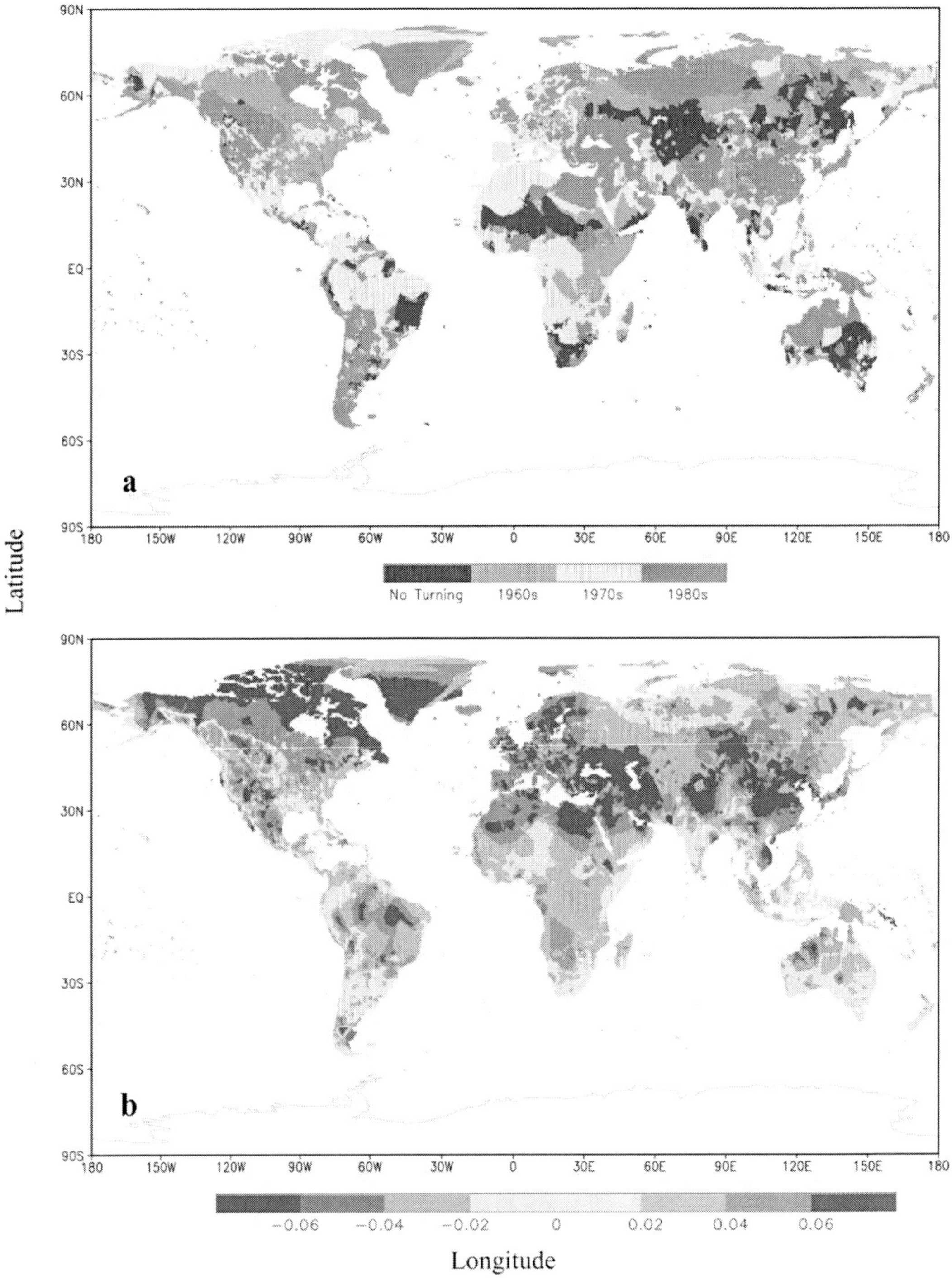

Figure 9. Occurring time of the latest inter-decadal trend turnings (ITT) of annual mean surface air temperature (SAT) in 1951～2002 (a), and linear trends after the breakpoint (b, °C/a).

Table 1. Ratio of the number of grids with various ITTs of SAT to the total number of grids in each region (%)

Time	No breakpoint		1960s		1970s		1980s	
Trend	<0	>0	<0	>0	<0	>0	<0	>0
ALA	0.00	4.03	0.00	43.28	0.00	36.01	13.13	3.54
CGI	0.00	0.00	0.00	16.01	0.03	30.07	2.23	51.67
WNA	0.00	3.63	0.00	11.00	0.04	20.89	42.78	21.67
CNA	0.00	0.00	0.00	55.28	0.00	27.01	12.08	5.62
ENA	0.00	0.16	0.00	62.73	0.00	20.24	4.66	12.21
CAM	0.00	4.84	0.00	14.83	0.10	59.03	13.12	8.07
NSA	0.00	8.07	0.11	2.98	0.04	74.88	5.67	8.25
CSA	0.00	19.99	1.54	2.23	0.45	37.76	13.15	24.89
SSA	0.00	9.25	1.80	3.61	4.55	5.64	58.15	17.01
NEU	0.00	5.08	0.00	29.27	0.00	23.34	0.29	42.02
SAM	0.00	0.20	0.00	1.54	0.00	57.25	0.04	40.97
SAH	0.00	12.39	0.47	10.38	0.44	42.83	0.03	33.47
WAF	0.00	30.88	0.00	9.33	0.00	47.44	9.33	3.00
EAF	0.00	12.08	0.00	56.48	0.16	20.19	7.29	3.79
SAF	0.00	19.02	0.00	10.24	0.00	51.10	14.42	5.21
NAS	0.00	22.97	0.00	31.77	0.00	11.26	25.97	8.03
CAS	0.00	27.61	0.00	8.20	0.00	19.31	0.07	44.81
EAS	0.00	12.24	0.00	16.72	0.00	16.29	0.55	54.20
SAS	0.00	14.56	0.00	20.95	0.00	27.60	11.73	25.16
SEA	0.00	9.75	0.00	15.97	0.00	39.66	24.22	10.40
WAU	0.00	13.37	0.00	0.19	11.73	9.71	57.06	7.94
EAU	0.00	44.98	0.00	0.41	15.60	4.55	22.32	12.14

1.4.3. Spatial Distribution of the ITT of Precipitation

We also studied ITTs of total annual precipitation for 1951–2002 by examining 67,359 terrestrial grids. Figure 10a shows the occurrence time of the last inter-decadal breakpoint of precipitation in each terrestrial grid. Most terrestrial grids registered an ITT in either the 1970s or 1980s. 25.7% of the terrestrial grids saw such changes in the 1970s and 62.5% in the 1980s. Linear trends after the breakpoint (Figure 10b) indicate that a decrease in precipitation (45.7% of the terrestrial grids) was mostly seen in North America, central South America, East Europe, the Sahara, Central and South Asia, northeastern East Asia, and east Australia. Along with SAT variability (Figure 9), a warming and increasingly dry trend has appeared in most grids in the past 20 or 30 years. The remaining 54.3% of the terrestrial grids experienced an ascending trend of precipitation after the last breakpoint, and this was most noticeable in northern South America, some parts of the West African coast, South China, Southeast Asia, and northwestern Australia. In the same manner, SAT variability has seen warmer and wetter conditions for most grids in northern South America, South China, and Southeast Asia. Northwestern Australia is a special case with most grids enjoying a decrease in SAT and an increase in precipitation. Table 2 presents the ratios of grids with different ITT patterns to the total number of grids in each region (an underlined number indicates the largest portion of ITT in a region).

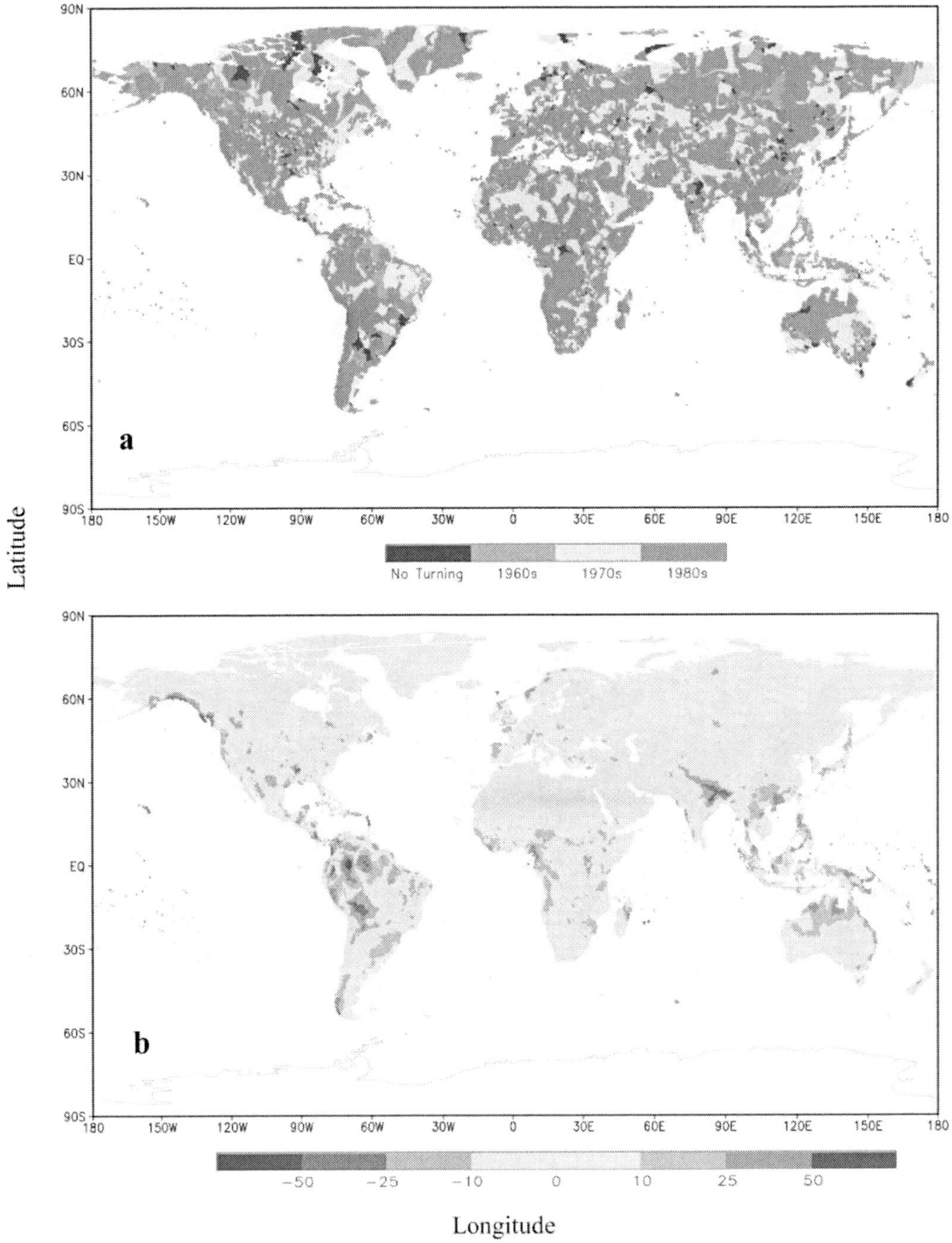

Figure 10. Occurring time of the latest ITT of annual total precipitation in 1951～2002 (a), and linear trends after the breakpoint (b, mm/a).

Table 2. Ratio of grids with various ITTs of precipitation to the total number of grids in each region (%)

Time	No breakpoint		1960s		1970s		1980s	
Trend	<0	>0	<0	>0	<0	>0	<0	>0
ALA	0.82	4.37	10.11	8.73	2.43	9.37	52.31	11.87
CGI	0.01	5.74	2.46	4.69	23.19	2.23	27.29	34.38
WNA	0.21	0.18	9.25	0.64	13.59	4.09	58.40	13.63
CNA	0.00	6.11	1.94	3.40	10.76	12.50	37.22	28.06
ENA	0.08	1.21	6.52	1.13	43.93	6.52	26.39	14.24
CAM	2.12	0.30	6.96	1.72	13.62	5.95	35.92	33.40
NSA	0.07	0.18	4.83	11.09	27.44	2.51	16.47	37.40
CSA	0.40	3.84	4.09	6.15	17.89	6.80	46.42	14.41
SSA	0.16	8.78	0.39	12.85	8.23	2.43	11.91	55.25
NEU	0.00	4.06	1.08	3.21	5.83	20.12	36.38	29.30
SEM	1.15	0.83	3.79	1.58	13.75	4.23	22.76	51.92
SAH	0.32	0.17	1.31	5.29	13.38	24.02	35.10	20.41
WAF	0.89	0.12	0.20	0.65	1.95	23.78	3.53	68.87
EAF	1.68	0.07	4.39	0.92	3.66	18.08	7.82	63.38
SAF	0.56	0.52	1.27	1.83	19.96	4.27	9.72	61.86
NAS	0.66	0.79	4.08	6.48	7.94	18.61	15.75	45.69
CAS	0.52	1.29	3.92	2.40	18.84	16.45	44.95	11.63
EAS	2.76	0.32	5.05	6.32	7.21	14.90	31.24	32.19
SAS	5.34	0.33	4.81	3.10	22.79	6.06	40.84	16.73
SEA	0.54	0.11	1.98	19.61	7.50	11.84	12.75	45.66
WAU	0.13	4.67	1.83	3.28	9.84	11.10	1.39	67.78
EAU	2.71	0.07	0.68	4.68	47.22	0.54	21.10	23.00

1.4.4. ITTs of Regional SAT and Precipitation

We analyzed ITTs of SAT and precipitation in the 22 regions for 1951–2002 (Figures 11-16) to understand climate changes in different regions in the last –50 years.

Figure 11 presents the results of ITTs in the six regions of North America. It is apparent that both ALA and ENA registered a breakpoint as early as 1965, corresponding to a change from decreasing to increasing temperatures, and two changes in precipitation corresponding to a change from increasingly wet to increasingly dry climates in both 1987 and 1980. ENA and ALA became warmer and drier in the early and late 1980s respectively. CAM had an ITT from decreasing to increasing temperatures in 1975 and an ITT from an increasingly wet to an increasingly dry climate in 1968, establishing a warmer and drier climate after the mid-1970s. The other three regions witnessed an ITT of SAT from decreasing to increasing temperatures and an ITT of precipitation from an increasingly wet to an increasingly dry climate, establishing a warmer and drier climate after the breakpoint. Overall, North America developed a noticeably warming and increasingly dry climatic trend.

The three regions in South America had different patterns for ITT. For example, NSA and CSA shared two ITTs of SAT, with the last ITT being from decreasing to increasing temperatures in 1980. In the context of precipitation, NSA and SSA shared a trend, with the last ITT taking place in 1987, from an increasingly dry to an increasingly wet climate, establishing a warm and wet climate in the 1980s. CSA had an ITT towards decreasing precipitation in 1982, establishing a warm and dry climate trend. SSA experienced a change in its temperature trend in 1987 towards decreasing temperature and has had an increasingly cold and wet climate since the late 1980s (Figure 12).

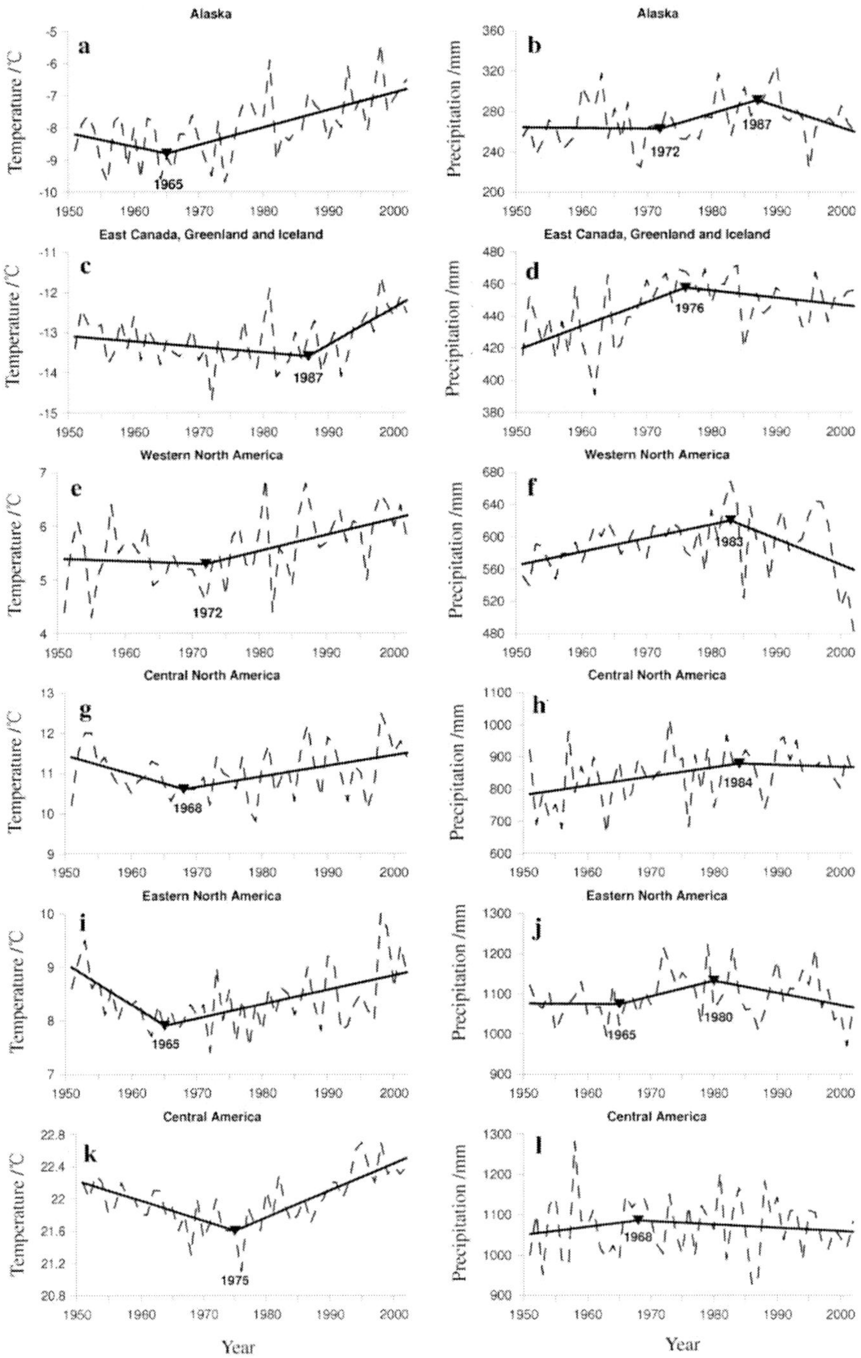

Figure 11. ITTs of SAT (a, c, e, g, i, k) and precipitation (b, d, f, h, j, l) in each region of North America during 1951–2002. Dotted lines represent annual variations, solid lines linear trends, and inverted triangles / numbers breakpoints.

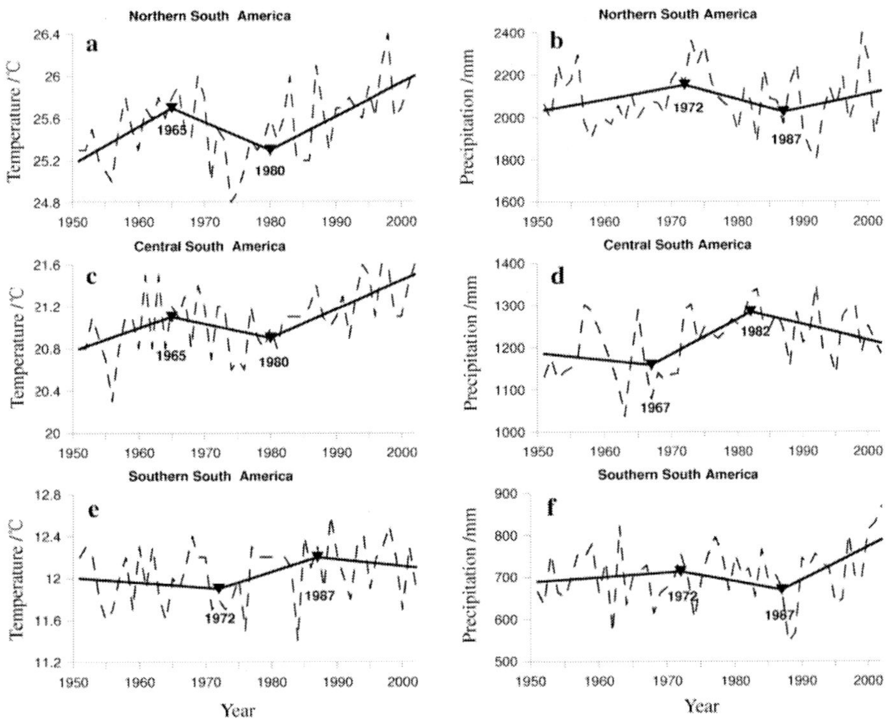

Figure 12. ITTs of SAT (a, c, e) and precipitation (b, d, f) in each region of South America during 1951–2002. Dotted lines represent annual variations, solid lines linear trends, and inverted triangles / numbers breakpoints.

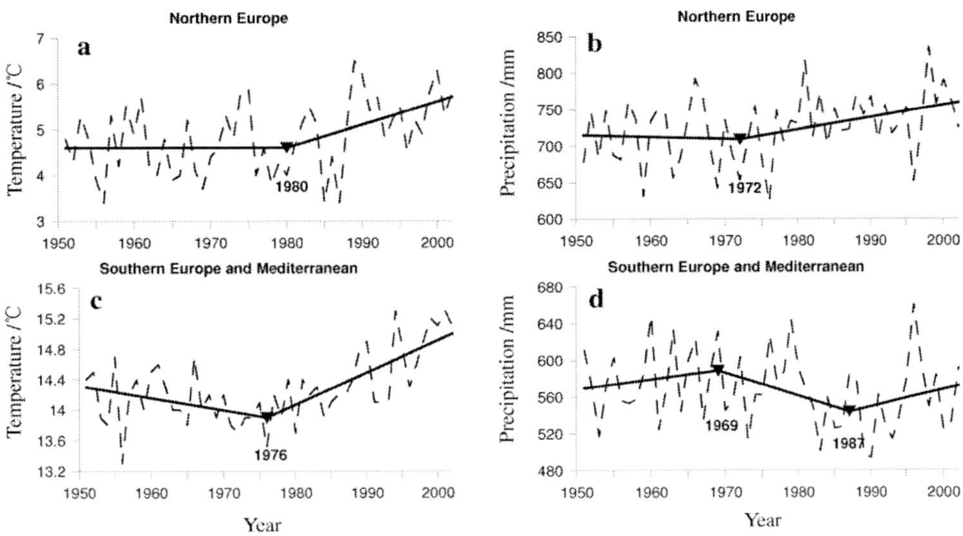

Figure 13. ITTs of SAT (a, c) and precipitation (b, d) in each region of Europe during 1951–2002. Dotted lines represent annual variations, solid lines linear trends, and inverted triangles / numbers breakpoints.

In Europe (Figure 13), NEU had two ITTs of annual mean SAT and total precipitation in 1980 and 1972, marking for an increasingly warm and wet climate since the early 1980s. SEM had only one ITT for SAT, in 1976, though it had two changes of the precipitation trend in 1969 and 1987. SEM became wetter after 1987, resulting in an increasingly warm and wet climate since the late 1980s.

Africa saw only one ITT of SAT in the period 1951–2002, with an ascending trend after the breakpoint. EAF recorded an ITT of SAT as early as in 1965, with the other three regions having the same ITT in the 1970s. Africa is quite different in the regional distribution of precipitation. For example, SAH began changing in 1971 from having decreasing precipitation to slightly increasing precipitation. WAF saw a change in precipitation in 1983 with an ascending trend. Both EAF and SAF had two ITTs towards increasing precipitation in the past –50 years, with the last breakpoints in 1984 and 1987 respectively. Overall, the African continent has experienced a warmer and wetter climate (Figure 14).

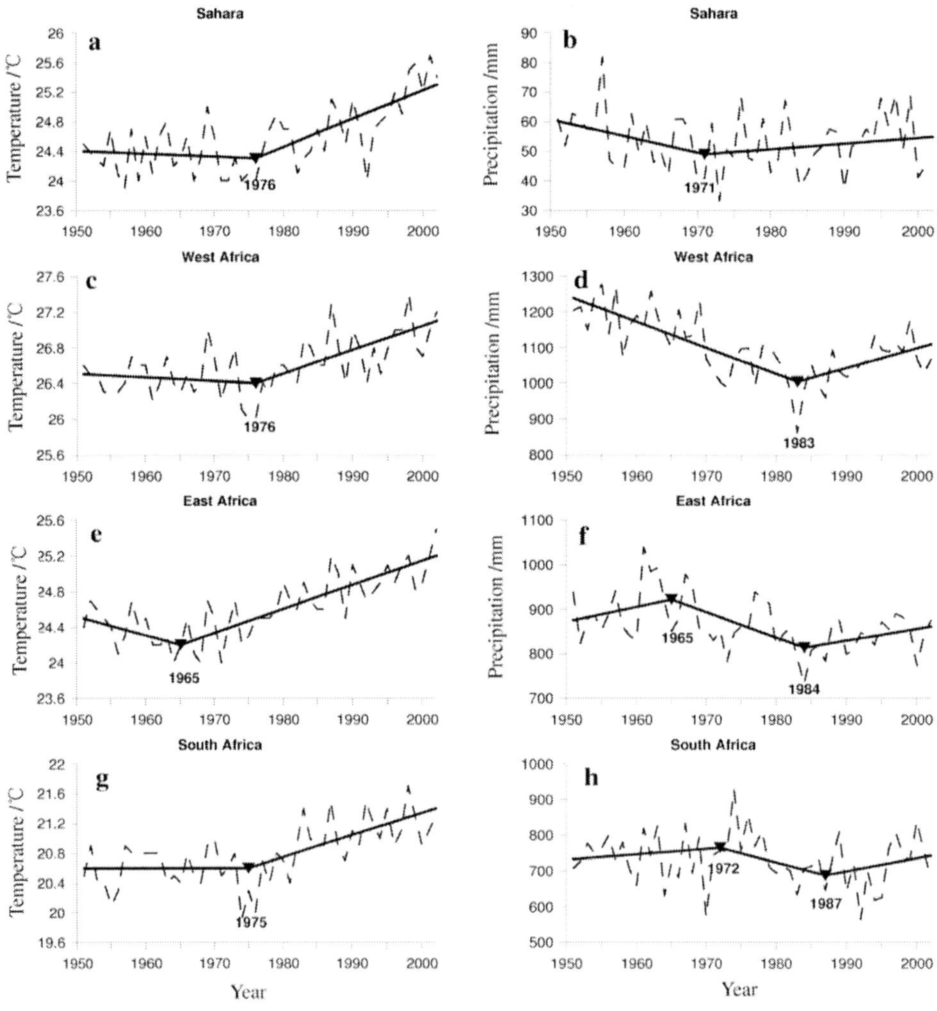

Figure 14. ITTs of SAT (a, c, e, g) and precipitation (b, d, f, h) in each region of Africa during 1951–2002. Dotted lines represent annual variations, solid lines linear trends, and inverted triangles / numbers breakpoints.

In Asia (Figure 15), NAS, SAS, and SEA registered an ITT of SAT in the period 1951–2002, with their breakpoints for a warmer climate occurring between the 1960s and 1970s. Meanwhile, both CAS and EAS saw two ITTs of SAT with an ascending trend before the 1960s, an ITT towards decreasing temperatures from the mid-1960s to the early 1980s, and a noticeable warming trend after the early 1980s. NAS and SEA shared the same change in annual total precipitation from decreasing precipitation before 1987 to increasing precipitation. EAS had an ITT around 1978, also towards an increasing trend. Both CAS and SAS experienced two ITTs of precipitation, with the last ITTs in 1987 and 1983 respectively. Both regions have established increasingly warm and dry climates since the 1980s.

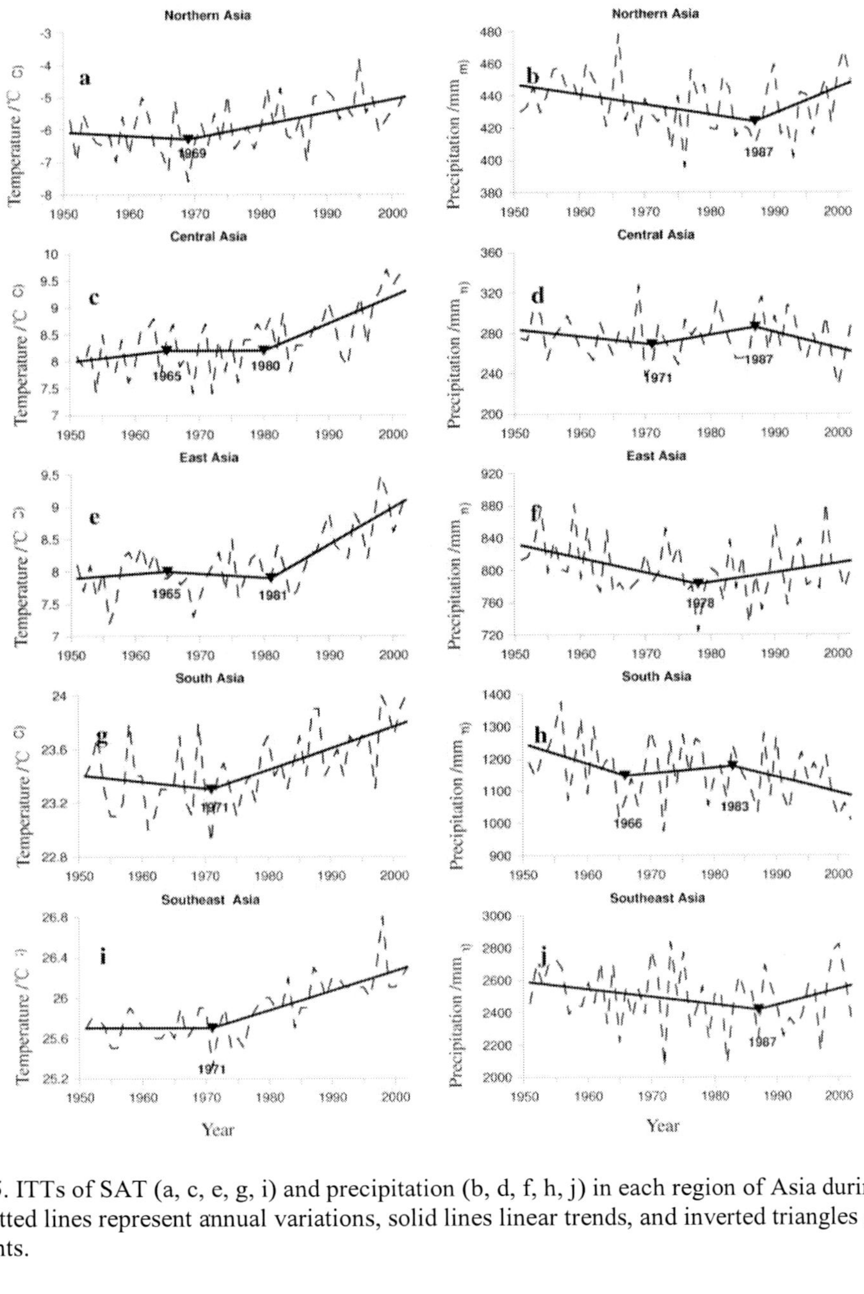

Figure 15. ITTs of SAT (a, c, e, g, i) and precipitation (b, d, f, h, j) in each region of Asia during 1951–2002. Dotted lines represent annual variations, solid lines linear trends, and inverted triangles / numbers breakpoints.

Figure 16 shows ITTs of SAT and precipitation in west and east Australia. West Australia changed from being increasingly warm to increasingly cold and from increasing dry to increasingly wet in 1987. West Australia has recorded an increasingly cold and wet climate since 1987. East Australia had no ITT of SAT in 1951–2002, having a linear warming trend, although there was an ITT of precipitation from increasingly wet to increasingly dry in 1974. East Australia has maintained an increasingly warm and dry climate since the mid-1970s.

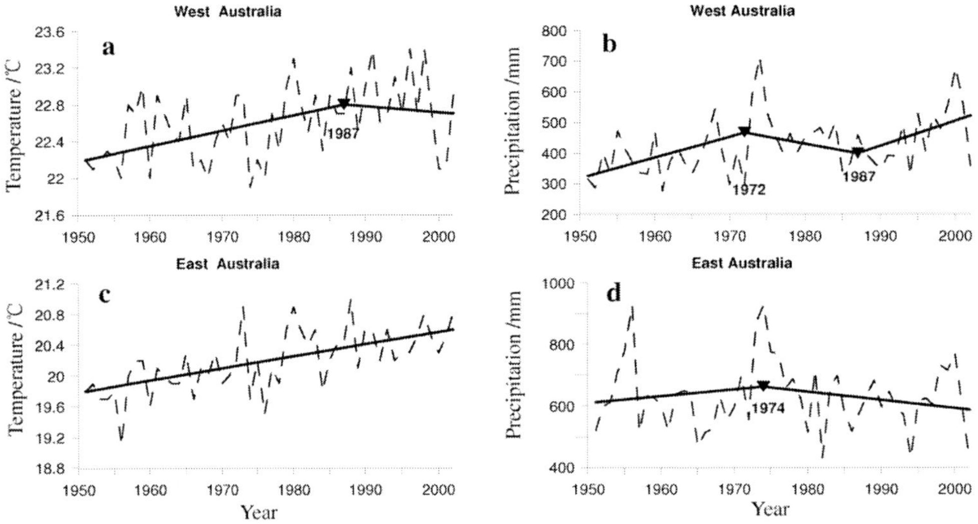

Figure 16. ITTs of SAT (a, c) and precipitation (b, d) in each region of Australia during 1951–2002. Dotted lines represent annual variations, solid lines linear trends, and inverted triangles / numbers breakpoints.

1.4.5. Conclusion

A grid-by-grid examination of ITTs of annual mean SAT and total precipitation for the 67,359 grid points in 1951–2002 and an analysis of the latest ITTs of SAT and precipitation in the period, in the context of both occurrence time of a breakpoint and the linear trend afterwards, have indicated that grids that have not had an ITT of annual mean SAT and have a warming trend and grids that have had a warming trend after an ITT of SAT account for up 85.7% of the terrestrial grids in the study. As a result, the warming trend has become highly significant in the global terrestrial system in the late 20th Century. Most terrestrial grids had an ITT of precipitation in either the 1970s or 1980s. Of these, the grids with an ITT in the 1970s accounted for 25.7% of all terrestrial grids, and the grids with an ITT in the 1980s accounted for 62.5%. Following the breakpoint, 45.7% of the terrestrial grids have had a decreasing trend in annual precipitation, with the remaining 54.3% having an increasing trend after the last ITT. Global terrestrial regions have basically become increasingly warm and dry or increasingly warm and wet.

An analysis of ITTs of regional mean SAT and precipitation in the 22 regions shows North America has become increasingly warm and dry after the last ITT. CSA, CAS, SAS, and EAU are also share the same climate trend. In Africa, four regions established increasingly warm and wet trends after the last ITT. NSA, NAS, EAS, and SEA in Europe moved towards to warmer and wetter climates, with most of Asia having the same trend. Both SSA and WAU moved towards colder and wetter climates after the last ITT.

2. DYNAMIC/THERMODYNAMIC STRUCTURES ASSOCIATED WITH A SERIES OF WINTER PRECIPITATION EVENTS OVER CHINA IN JANUARY 2008

2.1. Introduction

China's great snowstorms of 2008 have been identified as the third most memorable event in China during 2008, after Beijing's summer Olympic game and the devastating Sichuan earthquake. The snowstorms occurred on such a grand scale that 20 out of China's 34 provinces and special districts were affected by this disastrous event. According to Chinese Department of Civil and Internal Affairs, more than 100 million people suffered from these storms, and the economic loss totaled about 100 billion Chinese Yuan (about 15 billion U.S. dollars).

From early January to early February 2008, China experienced once-in-50-years (or once-in-100-years for some regions) snow and ice storms. These snow and ice storms were organized into four episodes, occurring during 10–16 January, 18–22 January, 25–29 January, and 31 January – 2 February. Most of southeast China was impacted by these large-scale storms. Figure 1a shows the distribution of accumulated total precipitation during this period (10 January–2 February) over China. Seven provinces exhibited total precipitation of over 100mm. If one uses a conservative snow-liquid conversion factor of 0.2 (NRCS, 2009), these provinces received more than 50 cm (~20 in) total snowfall during this period. Figure 17a also plots the distribution of snowfall and freezing rain areas. Most snow occurred in the north, which gradually transitioned into freezing rain in the south. The affected areas covered most of southeast and east-central China.

Figure 17b plots China's topography. From the southeast to the northwest, the altitude rises up in three steps. Such a topographic distribution leads to cold-air intrusion from Mongolia into northeastern China, then to southeastern China (shown as black arrows in Figure 17b). Based on these observations, we plotted regionally-averaged precipitation and temperature variations during 10 January–3 February, averaged over stations east of 105°E within a latitudinal band of 25°N-35°N (Figure 18a). Figure 18a shows that four significant precipitation episodes occurred during this period, accompanied by sustained low temperature, which exhibited a dramatic drop during the first precipitation episode.

The snow depth on 00UTC 28 January 2008 indicated that heavy snow was being accumulated in the region of the middle and lower reaches of the Yangtze River, among which parts of the Anhui and Jiangsu Provinces had reported snow depths of over 30 cm (Figure 18b). In Table 3, we listed averaged temperature and accumulated total precipitation and corresponding anomalies (calculated based on a climatology averaged from 1971-2000) for 13 provinces strongly affected by these snowstorms. One can see that most of these provinces showed an anomalous increase in accumulated total precipitation during this period. The situation was worsened by the anomalously low temperature which meant that a lot of the precipitation took the form of snow or ice accumulated on the ground. The snowstorms paralyzed both surface and air transportation, cut off electricity and other energy supply, and damaged agricultural crops. Even worse, these storms occurred right before the Chinese traditional New Year (5 February), when more than two hundred million people were on the move to their homes or on holiday trips. The storms trapped millions of people in railway stations, airports, and other transportation shelters.

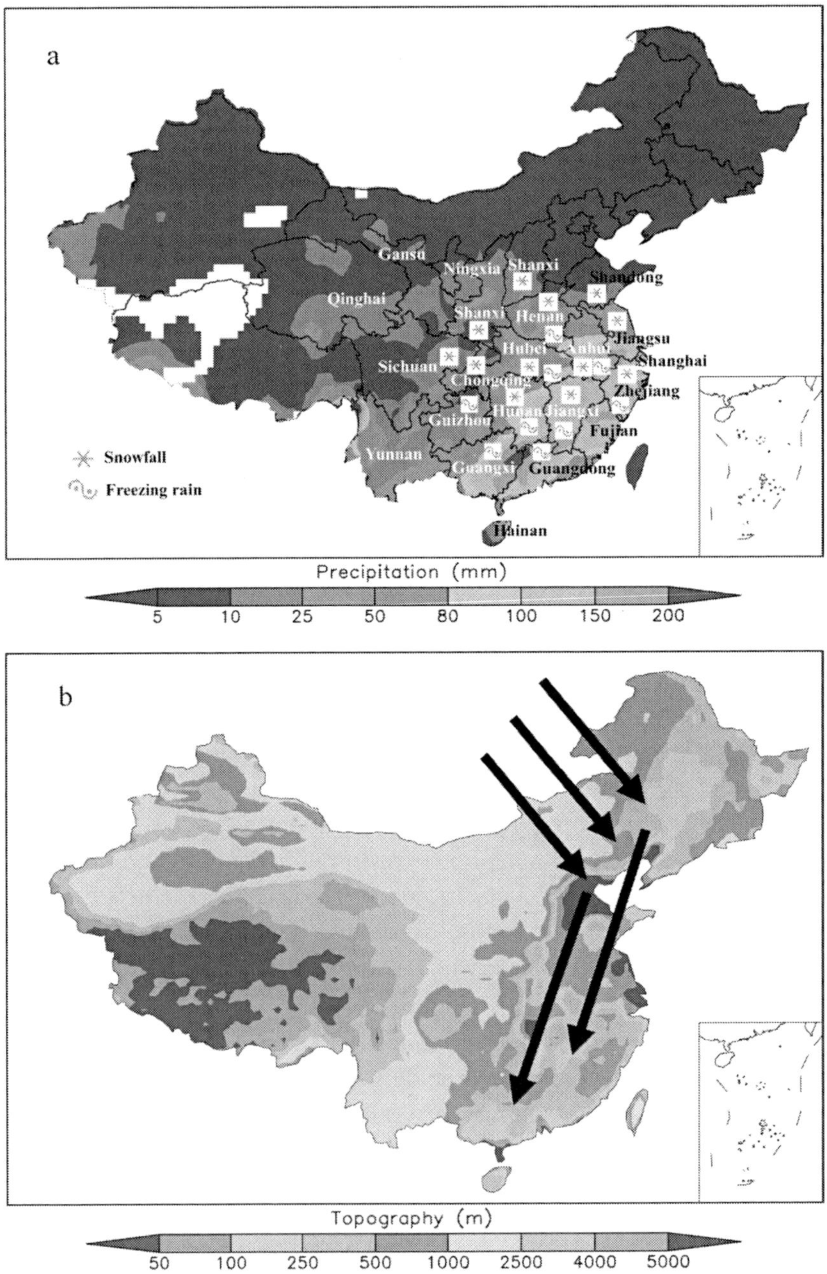

Figure 17. a) Distribution of accumulated total precipitation during China's great snowstorms of 2008 between 10 January to 2 February 2008 (unit: mm) as well as the major snowfall and freezing rain areas. Because snowfall is not a standard measurement in China, snow depth was unavailable. b) China topography (color-shaded) and cold air path during the snowstorms.

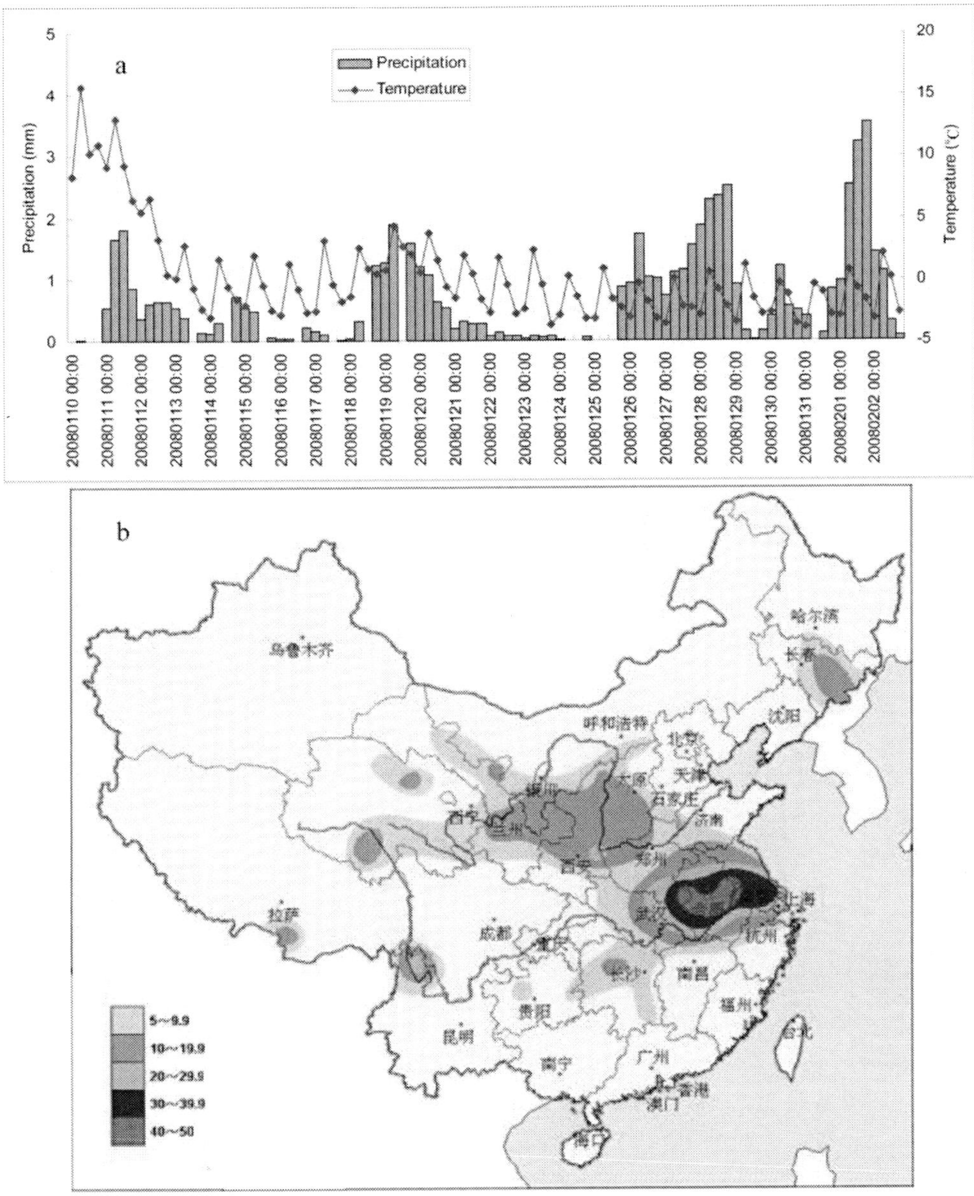

Figure 18. a) Time evolution of precipitation (area-averaged) and surface air temperature, averaged over a storm region (east of 105°E in a latitudinal belt of 25°N–35°N) during 10 January–3 February 2008 (time is in UTC). The precipitation includes stations that reported no precipitation. b) Distribution of snow depth on 00UTC 28 January 2008 (unit: cm).

Besides the unprecedented socio-economical impact brought by these storms, more than 50,000 Chinese meteorological professionals and scientists were under the media's spotlight–being questioned as to what they knew about how these extreme events happened on such a grand scale, and how much predictability they could provide for a future occurrence of storms like these. In recent years, many Chinese meteorologists and climatologists (Xu et al. 1999; Wang 2001; Li and Ding 2005; Shi et al. 2007) have indicated a weakening trend in the intensity of the East Asian winter monsoon since 1980. In response to the reduced monsoon intensity, China had experienced persistently warm winters for several years, and winter precipitation had also been decreasing (Wang and Ding 2006, Qian and Zhang 2007; Li et al. 2007a). There has been a large body of work concentrated on various mechanisms responsible for the recent warm winters of China. These studies pointed to the North Pacific Oscillation (NPO) (Wu and Wang 2002; Gong and Wang 2003; Gong et al. 2004; Chen and Kang 2006), El Niño and the Southern Oscillation (ENSO) (Tao and Zhang 1998; Gong and Wang 1999; He et al. 2007), the Siberian High (Qian et al. 2001; Gong et al. 2002), the Western Pacific teleconnection (Li et al. 2007b), and the East Asian jet stream (Yang et al. 2002; Mao et al. 2007); all of which may have influenced China's winter precipitation and temperature to some extent. On the other hand, Huang et al. (2007) suggested that the winter stationary planetary wave played an important role in the variation of the East Asian winter monsoon.

Table 3. Averaged temperature and total precipitation for major disastrous provinces and special districts during China's great snowstorms of 2008 (10 January–2 February 2008)

Factor / Region	Air temperature (°C)	Anomaly (°C)	Precipitation (mm)	Percentage of anomaly (%)
Shanghai	2.8	-0.9	106.6	123.9
Jiangsu	0.1	-1.5	72.3	137.8
Zhejiang	3.5	-1.0	114.1	149.7
Anhui	-1.2	-2.9	102.8	232.7
Jiangxi	1.4	-3.8	128.5	110.3
Henan	-2.9	-3.2	29.9	179.4
Hubei	-0.7	-3.8	54.9	128.8
Hunan	-0.5	-4.9	87.3	62.0
Chongqing	2.8	-3.1	16.3	-1.2
Guizhou	0.7	-4.1	24.3	14.6
Fujian	7.8	-1.0	117.6	165.5
Guangdong	10.3	-2.4	104.0	258.6
Guangxi	7.0	-4.2	111.2	269.4

Anomalies were calculated against region's climate records during the same period (the climatology is the average over 1971-2000).

Some synoptic studies for China winter snowstorms began to occur in the 1970's. Wang and Xu (1979) first investigated China's Inner Mongolia "77.10" extraordinary snowstorm. They proposed a "north ridge-south trough" anti-phased conceptual model for snowstorms in northern China. Gong and Li (2001) analyzed the relationship between Inner Mongolian snowstorms and low-level jets, and pointed out the importance of southerly low-level jets in the development of snowstorms in this region. Many other researchers also carried out diagnostic and modeling studies on snowstorms over the Tibetan Plateau (Deng et al. 2000;

Zhang and Chen 2000a, 2000b; Wang and Chen 2002). In recent years, there have been some case studies on abnormal snowstorms in the lower latitudes of southern China. For example, Zhu and Shou (1994) investigated secondary circulation and frontogenesis associated with a snowstorm that occurred in the Yangtze and Huaihe River basins. Yang et al. (2006) also conducted a modeling study of this storm using the MM5 model (NCAR-Penn State Mesoscale Model version 5). Nevertheless, knowledge of snowstorms in southern China is still quite limited, and understanding the physical principals that drive abnormal winter precipitation in lower-latitude southern China is still lacking.

During this grand snowstorm episode, the meteorological services from the central government to various local provinces all provided relatively accurate forecasts for each episode. Two- to five-day forecasts did indicate the storms and precipitation events would cover a large area of southeast China. However, being subject to the forecast lead time, the predictions from numerical weather forecast models were not able to give a sufficient warning for persistent low temperature and continued snowfall for more than 20 days. Therefore, the severity and persistence of these storms were underestimated. Furthermore, forecasters also realized that such extreme weather events were very improbable; even 50- and 100-year climate records did not include such persistent cold temperatures, consecutive heavy snowfalls, and ice storms for these areas.

The present study tried to provide a comprehensive understanding of how China's great snowstorms of 2008 occurred and developed, and depicted a picture of synoptic and storm-scale dynamic and thermodynamic structures. It needs to be clarified that the term "storms" used in this study are not a series of surface cyclones moving along a quasi-stationary frontal boundary through southeast China. The storms here are a prolonged period of warm advection in the mid-troposphere far to the north along a quasi-stationary frontal boundary. However, because of their large amount of snowfall, we refer to this series events as snowstorms.

2.2. Data

To investigate these snowstorms, we focused our data analyses over the period from 00UTC 10 January 2008 to 20UTC 2 February 2008. According to the four precipitation periods, we divided the data into four episodes: 10 January–16 January, 18 January–22 January, 25 January–29 January, and 31 January–2 February, for investigating each precipitation episode respectively. The composite of these four episodes provided a comprehensive understanding of physical mechanisms for the snow and ice storm's genesis, development, and reinforcement. These episodes also constituted four individual samples for comparison studies. We concentrated on the analyses of surface precipitation observations from 2519 weather stations around China, provided by the China Meteorological Administration (CMA). For other meteorological variables, we used the U.S. National Center for Environmental Prediction-National Center for Atmospheric Research reanalysis data (Kalnay et al. 1996), which provided geopotential height, meridional and zonal winds, humidity, and air temperature. The spatial resolution for this data is 2.5x2.5 lat-lon degree. We used six-hourly data for both surface precipitation observations and reanalysis data.

2.3. Synoptic Conditions and Large-Scale Atmospheric Circulations

Figure 19a shows averaged sea level pressure (SLP) and surface wind fields over the Asian region during 10 January–2 February 2008. Two main features that can be identified from this chart are the existence of the Siberian high (encompassed with blue rectangle) and prevailing northerly wind in eastern China and adjacent ocean (encompassed with orange rectangle). To understand the variation characteristics of the Siberian high during the snowstorm period, we plotted the latitude-time evolution diagram of SLP (contours) and its anomaly (color-shaded) averaged over the 70ºE–130ºE meridional belt (Figure 19b). One can see that this Siberian high-pressure system was sustained over the entire snowstorm period, and intensified during the four significant precipitation episodes. Compared with the climatology (averaged over 1971–2000), the SLP (between 40ºN–70ºN) exhibited positive anomalies, which indicates that the Siberian high had been anonymously strong during the period of the snowstorms. This situation was also accompanied by an intensification of corresponding northerly winds between 10°N–40°N in the longitude band of 110ºE~130ºE (Figure 3c). The sustained negative anomalies in the meridional wind indicated that there had been stronger northerly winds in eastern China and the adjacent ocean than normal. The anomalous northerly winds brought significant cold air into eastern China.

Figures 20a-d are composites of 500-hPa geopotential height and its anomaly fields over the Asian region, corresponding to the four significant precipitation periods. There existed a blocking ridge north of China during all of these cases. The corresponding blocking centers are respectively at (90°E, 62N), (70°E, 55°N), (85°E, 55°N), and (70°E, 48°N). In the mean time, the western Pacific subtropical high (WPSH) between 10°N–25°N also showed above-normal strength. During the four precipitation episodes, the positive-negative-positive patterns of geopotential anomalies were sustained. While the variations of the northern blocking ridge affected cold air moving southward, the variations of the WPSH controlled warm-moist air transport northward. In normal winter years, a large-scale cold-air intrusion from the north often causes a temperature drop in a large area in China. When combined with a supply of moisture from the south, a winter weather system often brings snows in northern China and rains in southern China. Such a winter system typically lasts between five and seven days. With the breakdown of the northern blocking ridge and weakening of the WPSH, cold-dry northerly wind and warm-moist southerly flow were weakened. The typical large-scale circulation pattern for winter precipitation in eastern China would either move out to sea or dissipate entirely. However, China's great snowstorms of 2008 occurred during a period of 20 days with some heavy precipitation with four distinct episodes of enhanced precipitation, each lasting a little over five days. From the large-scale synoptic weather patterns indicated in Figures 3-4, we observed that in the present case, the typical large-scale circulation pattern formed, but never moved on or dissipated. There was a strong negative height anomaly over Japan and north of the WPSH which would result in the stronger than normal jet over northeast China that will be discussed later in the paper. This would act to keep the entrance region of the jet persisted over eastern China.

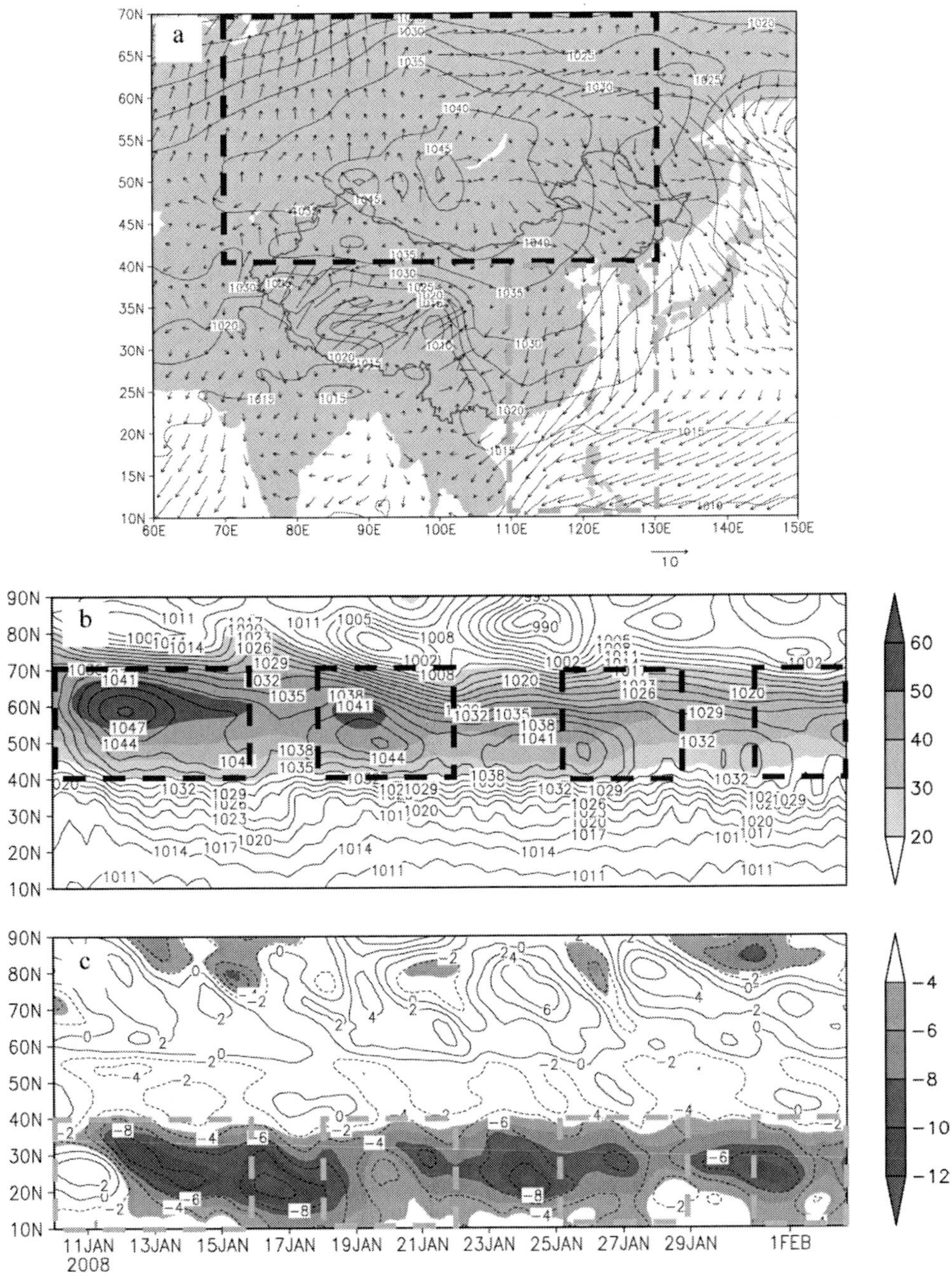

Figure 19. a) Horizontal distribution of sea level pressure (SLP, contour interval is 5 hPa) and surface wind (m/s) fields averaged over the period 10 Jan to 2 Feb over the Asian region (the red is border line of China); b) latitude-time evolution of the SLP (contour interval is 3 hPa) averaged over 70°E–130°E and its anomaly (color-shaded, unit: hPa) ; c) Latitude-time evolution of the surface meridional wind (contour interval is 2 m/s) averaged over 110°E–130°E and its anomaly (color-shaded, units: m/s). The blue dashed rectangle in b) and orange dashed rectangle in c) indicate the time and location of the precipitation. The climate is the mean for 1971~2000.

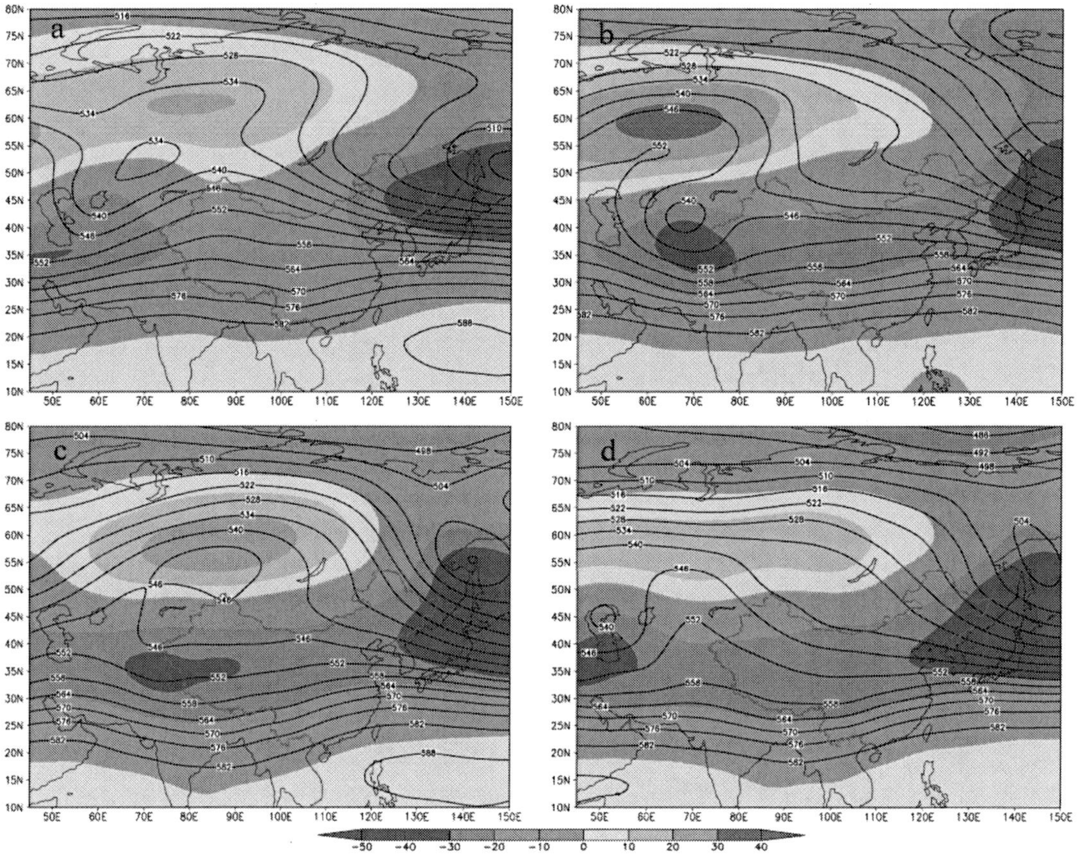

Figure 20. Averaged geopotential height (contour, interval is 60gpm) and its anomaly (color-shaded, in unit: 10gpm) at 500 hPa pressure level during a) 10 January–16 January, b) 18 January–22 January, c) 25 January–9 January, and d) 31 January–2 February 2008.

In Figure 21a, we plot the 700 hPa heights and temperatures, averaged over the period of January 10–February 2, 2008 (24 days). The geopotential height (solid-black) presented a ridge in northwestern China and a trough over central China. This pressure pattern caused flows to confluent downstream (indicated by two arrows). There was also a thermal trough (marked by thick dashed line) in the vicinity of the flow confluence region in southeast China (about 20–35°N and 105–120°E). The northern branch of the convergent flow brought in cold-dry air, while the southern branch brought in warm-moist air. The development of the confluence zone led to an upgliding of warm-moist air over the cold-dry air, thus heavy snows occurred along the confluence region. All these features were very similar to a study (Figure 21b) of the 1977 snowstorm over Inner Mongolia by Wang and Xu (1979). Their finding indicates that the "north ridge-south trough" anti-phase pressure field coupled with confluence may characterize the typical synoptic conditions for snowstorm in east China. From the latitude-time evolution diagram (Figure 22) for 700-hPa geopotential height (averaged over 70°E–110°E), one can also see that the signature of the "north ridge-south trough" corresponds to each precipitation process. In each episode the trough (dashed ellipse) was pushed by the strong blocking ridge (solid ellipse) from the north deep into south below 30°N. As noticed previously, the presence of a strong subtropical high off the southeast coast (at 20°N) of China might contribute to the persistence of snowstorms. The subtropical high not only sustained the precipitation by supplying a large amount of warm-moist water vapor to southeastern China, but also blocked the troughs from moving out to sea.

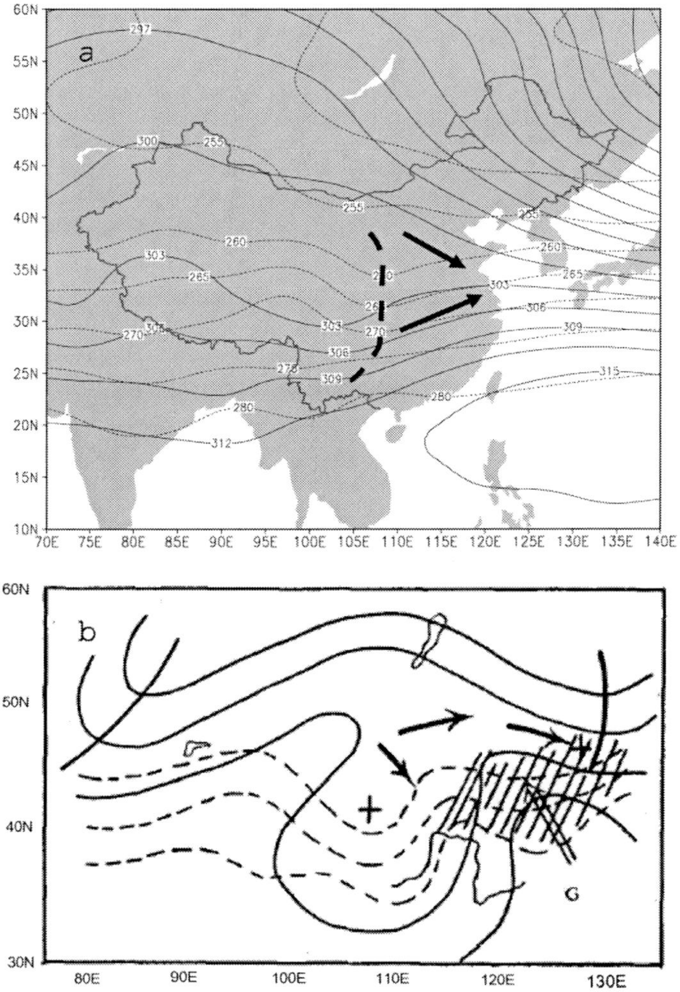

Figure 21. a) Averaged (over 10 January–2 February 2008) 700hPa geopotential height (solid, interval 30 m) and temperature (dash, interval 5K). b) The schematic map at 700hPa for a big snowstorm in 1977 in Inner Mongolia, China from a previous study by Wang and Xu (1979).

To examine wind, temperature, and precipitation fields in the low-level atmosphere, we further plotted these fields at 925 (left column) and 850 hPa (right column in Figure 23) for each episode from top to bottom, respectively. During each significant precipitation process, a cold tongue (or thermal trough) was evident in the lower levels of atmosphere, just west of the heavy precipitation (color-shaded). The cold tongue was stronger at lower levels than above, because at 925 hPa, the dominant wind direction (arrow) in this region was northeasterly (Figure 23a, c, e, g), while at 850 hPa, flows were mostly southwesterly (Figure 23b, d, f, and h). The 925 hPa and 850 hPa field also seem to show some evidence of cold air damming especially with the strong ageostrophic northeast flow at 925 hPa.

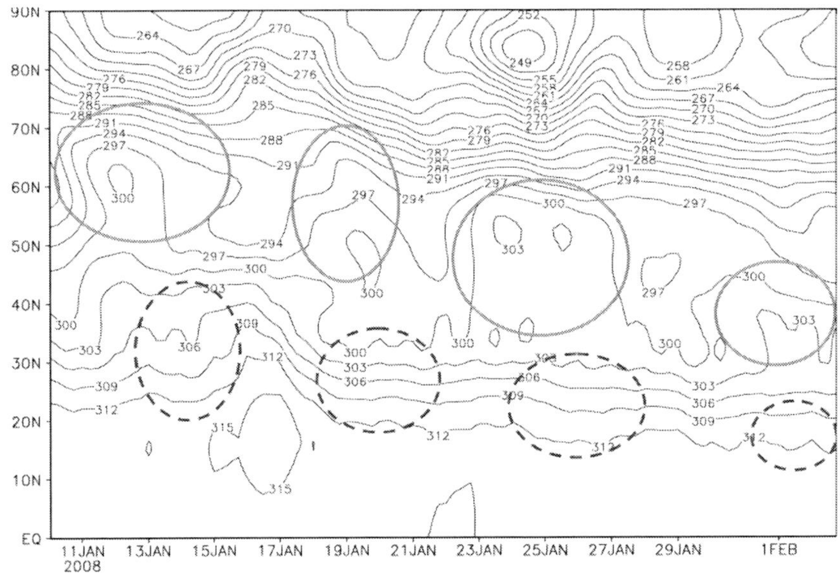

Figure 22. Latitude-time evolution of averaged (over 70°E-110°E) geopotential height field at 700-hPa pressure level during 10 January–3 February 2008 (unit: 10gpm). Solid circles outline the locations of ridges, and the dashed circles outline troughs.

To better understand the variations of synoptic forcing and precipitation processes, we conducted detailed analyses for a particular storm episode. The third precipitation episode (25–29 January 2008) was selected for this study. Figure 24 shows the SLP over the north of China every 24 h. One can see that the Siberian high intensified from 25 to 26 January (Figures 24a and b). From 27 to 29 January (Figures 24c-e), the Siberian high pressure decreased slightly, but it persisted and continuously expanded southward. Corresponding to this, the northerly winds (on 925-hPa level) presented strengthening –weakening - strengthening variation, but were persistent. At 500 hPa (Figure 25), the Ural-blocking high and a weak anti-phased trough established on 26 January persisted throughout the storm episode. During the same period, the WPSH also showed a strong signal (see the 588 dm contour), which expanded westward on 26 and 27 January. It began to weaken and retreat to the east on 28–29 January. Associated with the setup of the WPSH and upper-level trough, strong water vapor transport occurred from the Bay of Bengal and the South China Sea, seen at 700hPa by the wind (vectors) and specific humidity (color-shaded) analysis in Figure 26. On 29 January, both water vapor transport and specific humidity displayed significant decreases. Figure 27 shows that heavy precipitation occurred on 25 January (color-shaded). It weakened on the 27th, regained strength on the 28th, and began to dissipate on the 29th of January. The contours plotted on Figure 27 are 200-hPa zonal winds, which show the development of the upper-level jet streams. Figure 27a, d shows that the greatest precipitation falls in the right entrance region of the jet maximum to the east of China associated with the large scale confluence. The precipitation weakens in Figure 27b, c when the jet streak propagates through southern China and the precipitation region is located in the right exit region. Figures 27 b, c also suggest that the primary dynamic mechanism for the precipitation is not the jet-streak circulations. Rather, the precipitation is enhanced when it is in a favorable location with respect to the jet. This indicates that the upper-level divergence in right entrance region of a jet stream poses a "pumping effect" to maintain an upward motion, thus promoting precipitation in the region. As the jet core moved out of China (Figures 27 e), the precipitation began to dissipate.

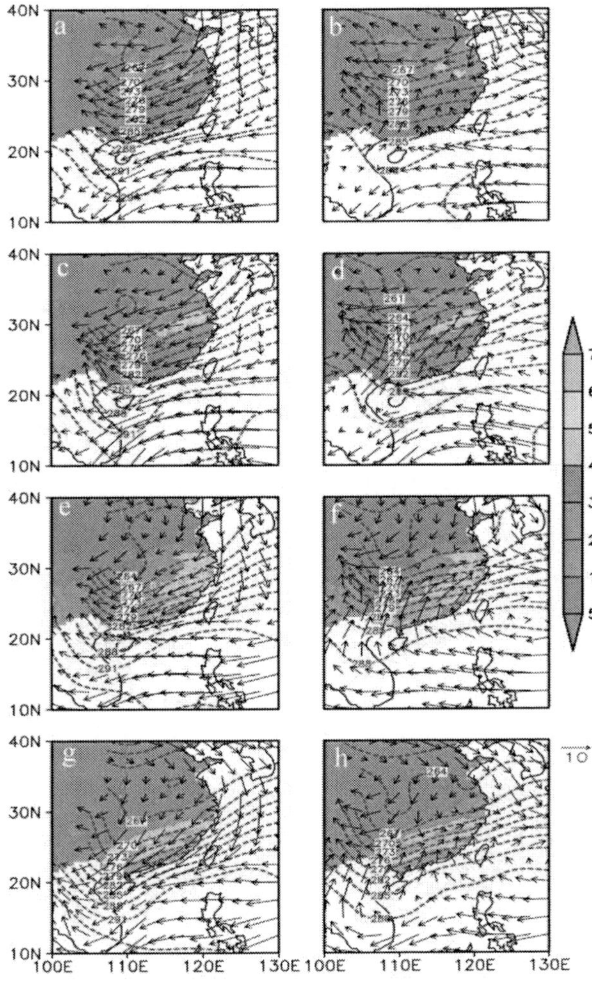

Figure 23. Averaged temperature (red dashed, interval 3K), wind vector (arrow, m/s), and total precipitation (color-shaded, mm) for 925 hPa (left column) and 850 hPa (right column) pressure levels during 10 January–16 January (a, b), 18 January–22 January (c, d), 25 January-29 January (e, f), and 31 January-2 February (g, h) 2008.

Comparing these results with U.S. northeast snowstorms (Kocin and Uccellini 2004; denoted as KU hereafter), we find that many of these main features are strikingly similar between snowstorms on the two continents. For example, the large-scale pattern of a dominant upper-level trough coupled with a weak ridge to its north resulting in confluent flows east of the pressure centers was also found in Figure 7-4 of KU. In addition, the high-pressure system in the north leading to cold-air damming in Figures 7-4, 7-6, and 8-1 of KU is conceptually the same as the cold tongue intrusion from the north in the China snowstorms analyzed in this study. However, a distinctive feature shown in the China snowstorms was the presence of a subtropical high over its coast. As we discussed above, this subtropical high might play an important role in blocking the snowstorms exiting to the sea, which explains why the snowstorms lasted on and off for almost a month inland. Of course the subtropical high also helped to advect warm and moist air northward. In the U.S., this subtropical-high blocking effect was typically not seen. Furthermore, differences in geographic features from the two countries may also result in differences in snowstorms, e.g., storm locations. China's great snowstorms of 2008 occurred along the southeast coast of China, between 20–35°N, while U.S. snowstorms typically occurred in the northeast coast, around 35–45 ° N. In China, the world largest and highest plateau and mountains, the Tibetan Plateau and Himalayas, are located upstream of the snowstorms, directly to the west. The plateau blocks water vapor transported from low-latitude oceans to northern China. Instead, it deflects the moisture transport to the east, which promotes a storm development along the southeast coast of China.

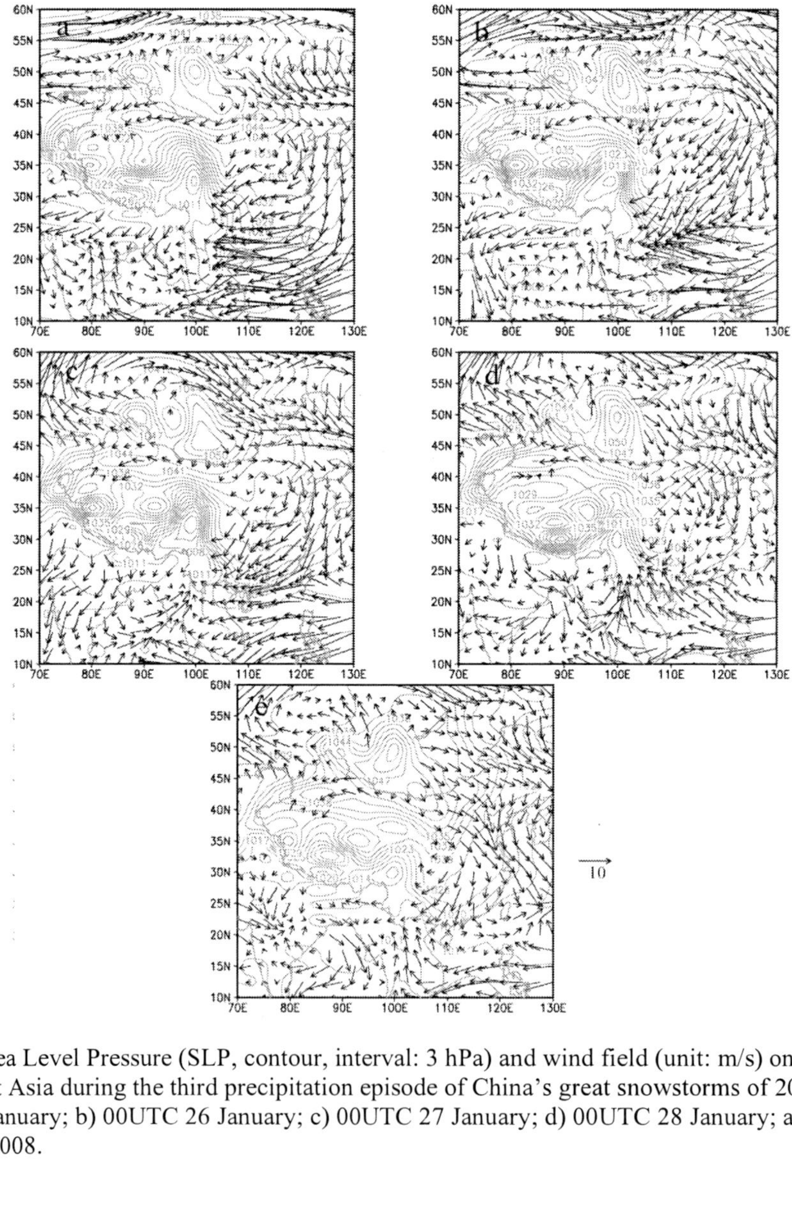

Figure 24. Sea Level Pressure (SLP, contour, interval: 3 hPa) and wind field (unit: m/s) on 925 hPa over the East Asia during the third precipitation episode of China's great snowstorms of 2008. a) 00UTC 25 January; b) 00UTC 26 January; c) 00UTC 27 January; d) 00UTC 28 January; and e) 00UTC 29 January 2008.

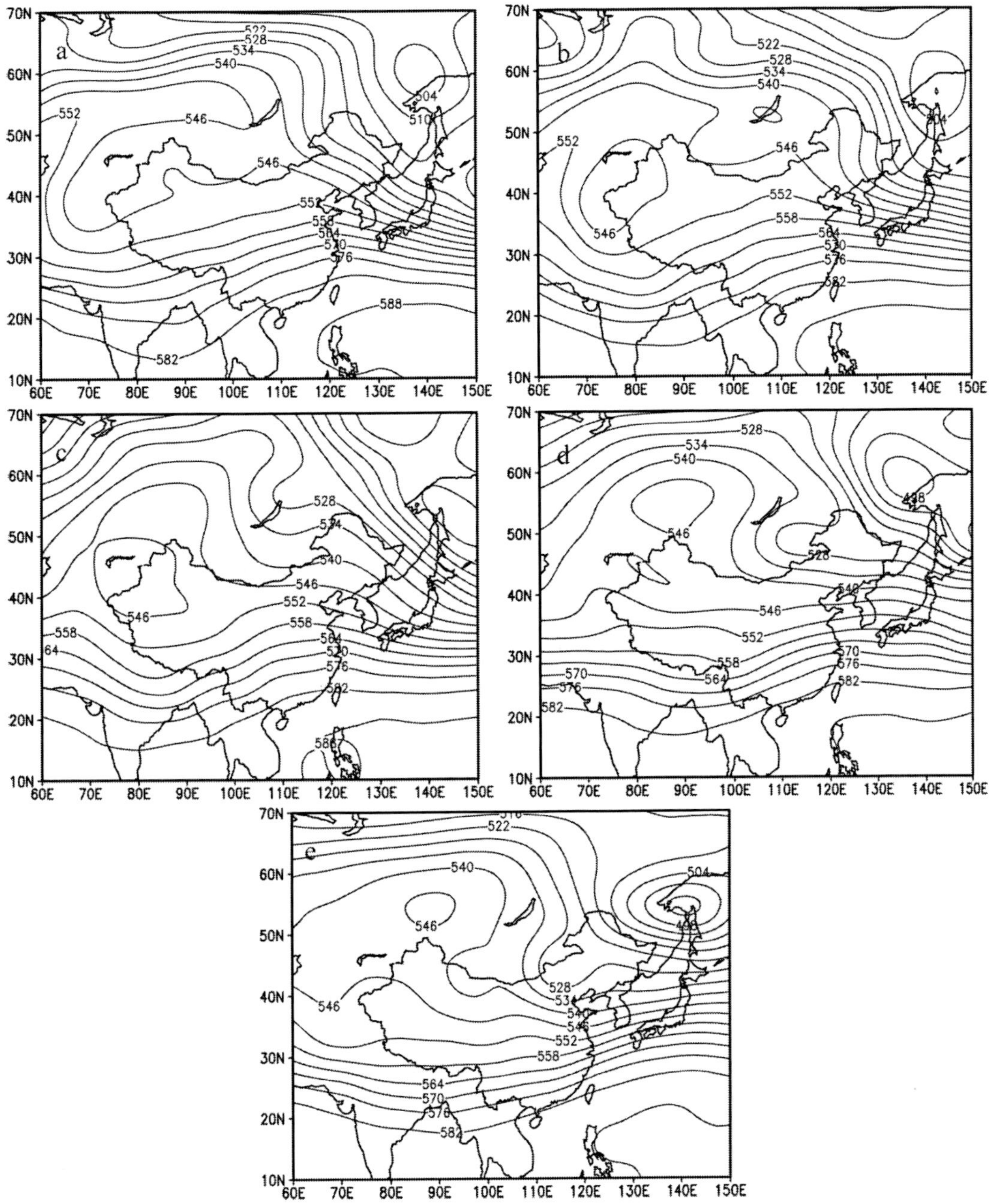

Figure 25. Geopotential height at 500 hPa over Asian region during the third precipitation episode of China's great snowstorms of 2008 (contour interval: 60gpm). a) 00UTC 25 January; b) 00UTC 26 January; c) 00UTC 27 January; d) 00UTC 28 January; and e) 00UTC 29 January 2008.

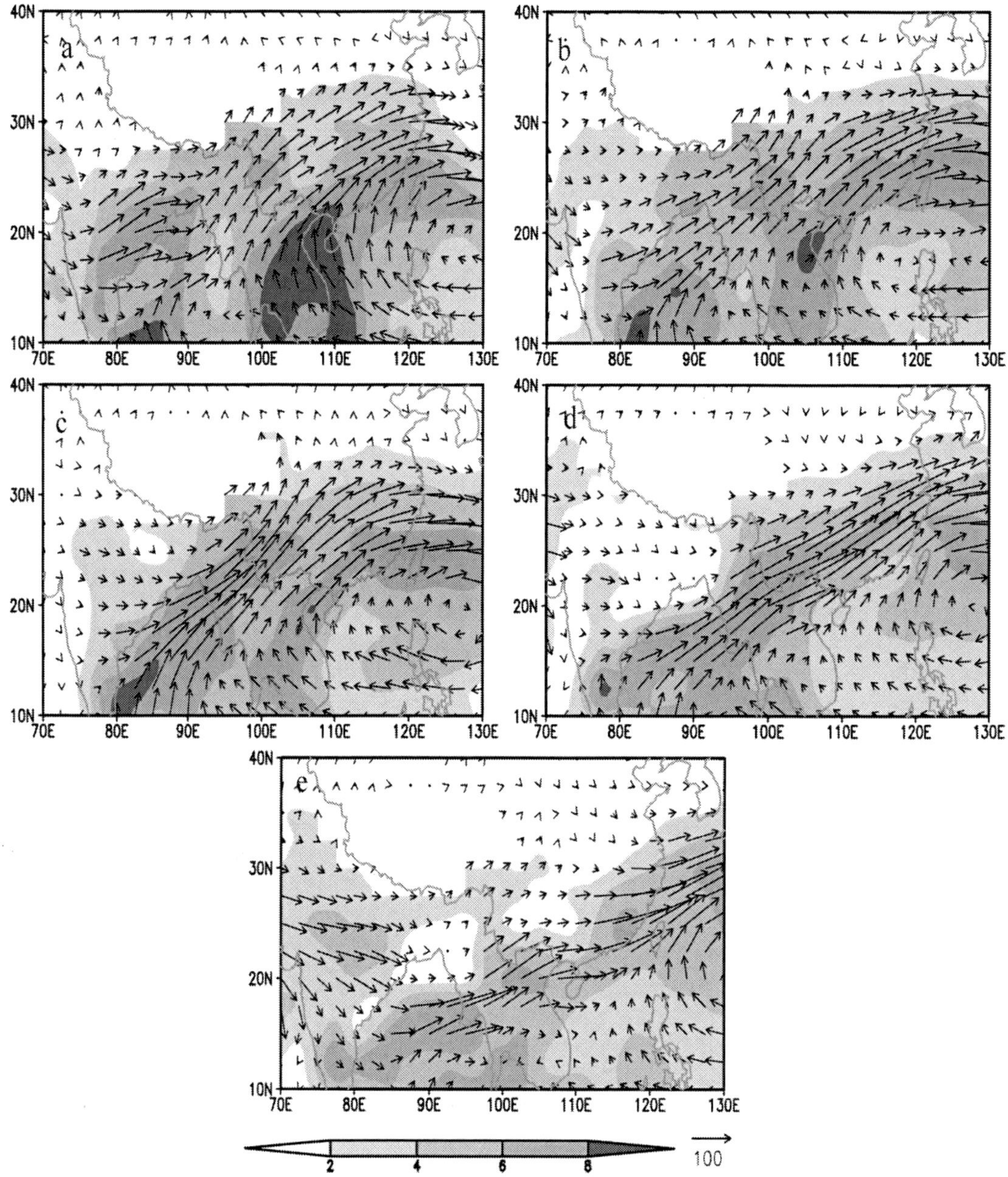

Figure 26. Water vapor flux (vector; unit: g.m.kg-1.s-1) and specific humidity (color-shaded; unit: g.kg-1) on 700 hPa over southern China during the third precipitation episode of China's great snowstorms of 2008. a) 00UTC 25 January; b) 00UTC 26 January; c) 00UTC 27 January; d) 00UTC 28 January; and e) 00UTC 29 January 2008.

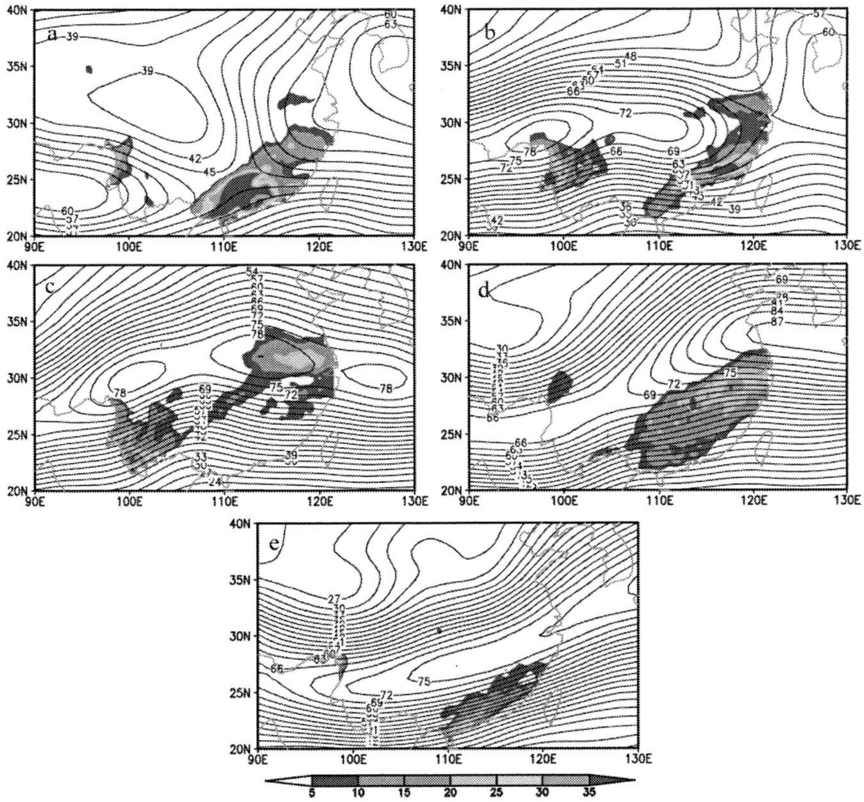

Figure 27. 24-h accumulated surface precipitation (color-shaded; unit: mm) for the day labeled below and zonal winds (contour; interval is 3 m/s) on 200 hPa over southeastern China during the third precipitation episode of China's great snowstorms of 2008. a) 00UTC 25 January; b) 00UTC 26 January; c) 00UTC 27 January; d) 00UTC 28 January; and e) 00UTC 29 January 2008.

2.4. Vertical Wind Shear, Upper-Level Jet, and Temperature/Moisture Inversion

Note that all four episodes of snowstorms essentially occurred in the same large areas shown in Figure 18a. All features discussed above are common to all four snowstorms that occurred during this period. The only difference is the varying strength of these features from one storm to another. Thus, we can examine the flow fields by taking an average of the four episodes. We plotted averaged meridional wind (contours) and its anomaly (color-shaded) in a south-north vertical cross section (Figure 28a) and averaged zonal wind (contours) and its anomaly (color-shaded) in a west-east vertical cross section (Figure 28b). It can be seen that in the snowy region (blue-shaded), vertical wind shear was evident in both meridional and zonal wind components at lower levels. For example, for the meridional wind in Figure 28a, southerly wind prevailed above 850 hPa, while below 850 hPa, northerly wind prevailed. The anomaly showed negative-positive coupling from the bottom up, indicating a stronger southerly wind at upper levels and a stronger northerly wind at lower levels in comparison to climatology. Similarly, for the zonal wind in Figure 28b, westerly wind prevailed above 850 hPa and a very strong westerly jet was located at the upper troposphere (around 200 hPa) with a positive anomaly, while below 850 hPa, easterly wind prevailed with a negative anomaly. The low-level northeasterly wind brought in cold-dry air, which undercut the warm-moist air brought by the southwesterly wind at middle levels.

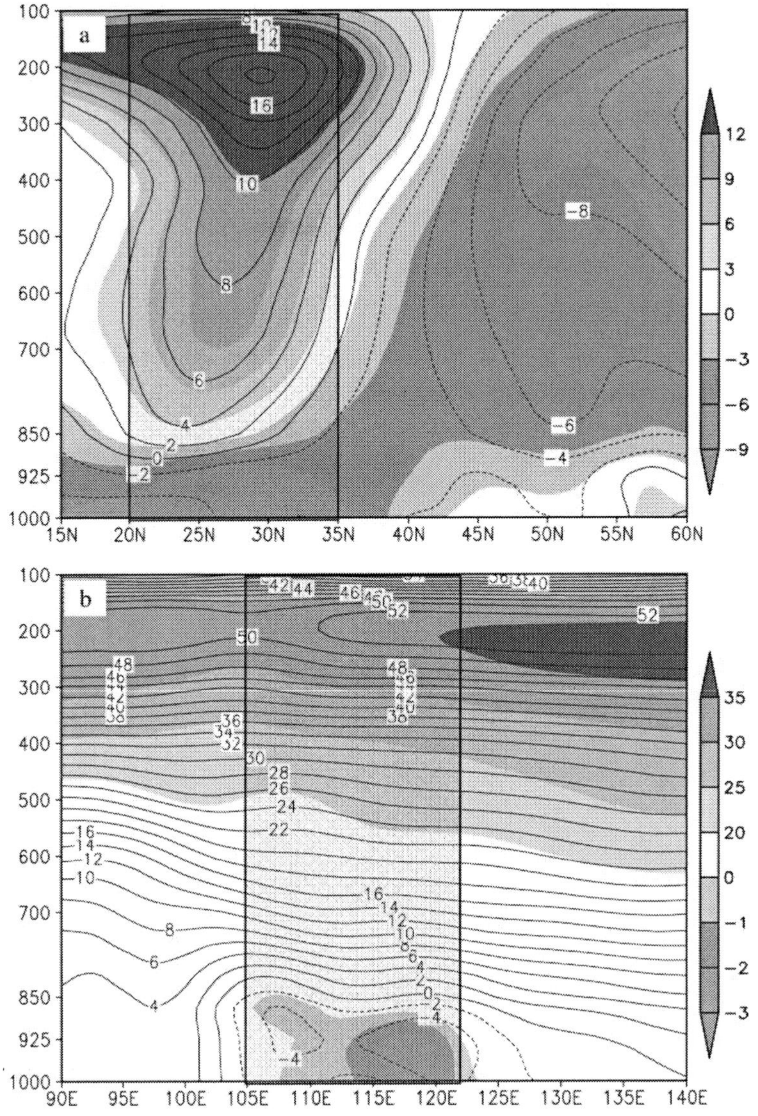

Figure 28. Averaged (over 10 January–2 February 2008) a) meridional wind (contour, interval is 2 m/s) and its anomaly (shaded, unit: m/s) in the south-north vertical cross section along 110°–120°E longitudes; and b) zonal wind (contour, interval is 2 m/s) and its anomaly (shaded, unit: m/s) in the west-east vertical cross section along 22.5°–32.5°N latitudes. Blue-shaded regions indicate the regions of snowstorms.

The vertical wind shear also suggests a thermodynamic structure of cold-air intrusion at low levels and warm-air crossover at middle levels. During the period of persistent freezing rain and snow, the third episode (25–29 January 2008) was the most severe snowstorm. Figures 29 and 30 show the vertical cross sections of temperature and specific humidity coupled with vertical atmospheric circulations in this episode. In the south-north vertical cross section (Figure 29a), the isotherms (color-shaded) between 925–700-hPa levels tilted towards the north with increasing altitude, indicating a temperature inversion and warm advection in the region of heavy precipitation (Figure 29c). In the west-east vertical cross section (Figure 29b), a cold-air center existed in the low levels around 110ºE longitude. This cold-air center is a slab view of the cold tongue that we saw in Figure 23, representing a cold-air intrusion from north to south in the lower atmosphere. On the east side of this cold-air center, precipitation occurred over a large area (Figure 29d). Over this region between the 925–700-hPa levels, the isotherms (color-shaded in Figure 29b) tilted westward with the increase of altitude, which also indicates a temperature inversion.

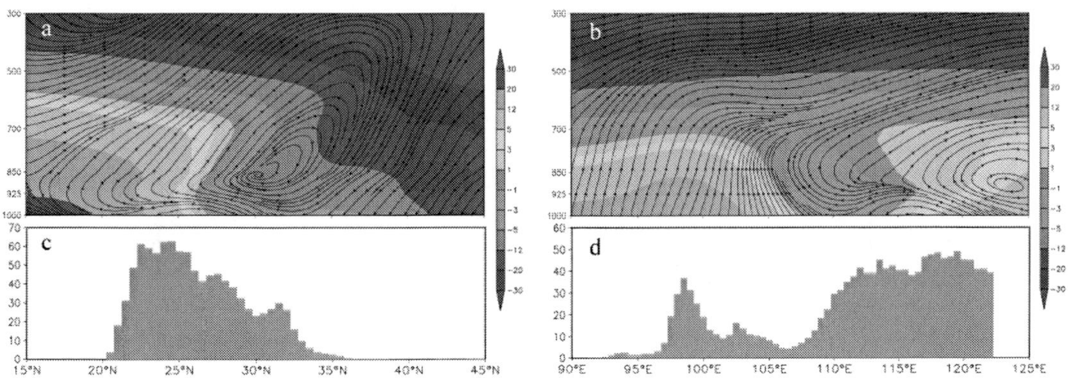

Figure 29. Averaged vertical circulations (arrow streamlines) and temperature (color-shaded, °C) a) in the south-north vertical cross section along 110°–120°E; b) in the west-east vertical cross section along 22.5–32.5°N during the period of 25 – 29 January 2008 ; c) and d) are corresponding total precipitation along these vertical sections (unit: mm).

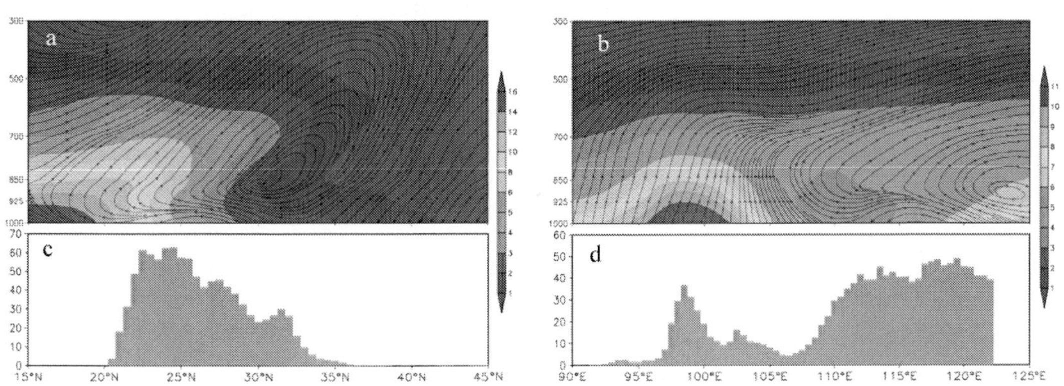

Figure 30. Averaged vertical circulations (arrow streamlines) and specific humidity (color-shaded, g.kg-1) a) in the south-north vertical cross section along 110°–120°E; b) in the west-east vertical cross section along 22.5°–32.5°N during the period of 25 – 29 January 2008; c) and d) are corresponding total precipitation along these vertical sections (unit: mm).

Very similar vertical structure is also found in the humidity field. In the south-north vertical cross section (Figure 30a), the specific humidity (color-shaded) tilted towards the north with increasing altitude between 925–700-hPa levels, right above the snowfall area (Figure 30c). This indicated a moisture inversion in the region of heavy precipitation as well. In the west-east vertical cross section (Figures 30b and 30d), a dry-air center existed in the low levels around 110°E longitude, in association with the cold-air intrusion from north to south at lower levels. On the east side of this cold-air center and between 925–700-hPa levels, the specific humidity contours tilted towards the north with increasing altitude, also depicting a moisture inversion. The moisture inversion manifested the moisture was lifted over the cold dome of air, presumably condensing in the upper levels to produce the snow/frozen precipitation. The other three precipitation episodes showed similar temperature and moisture inversion in the vertical.

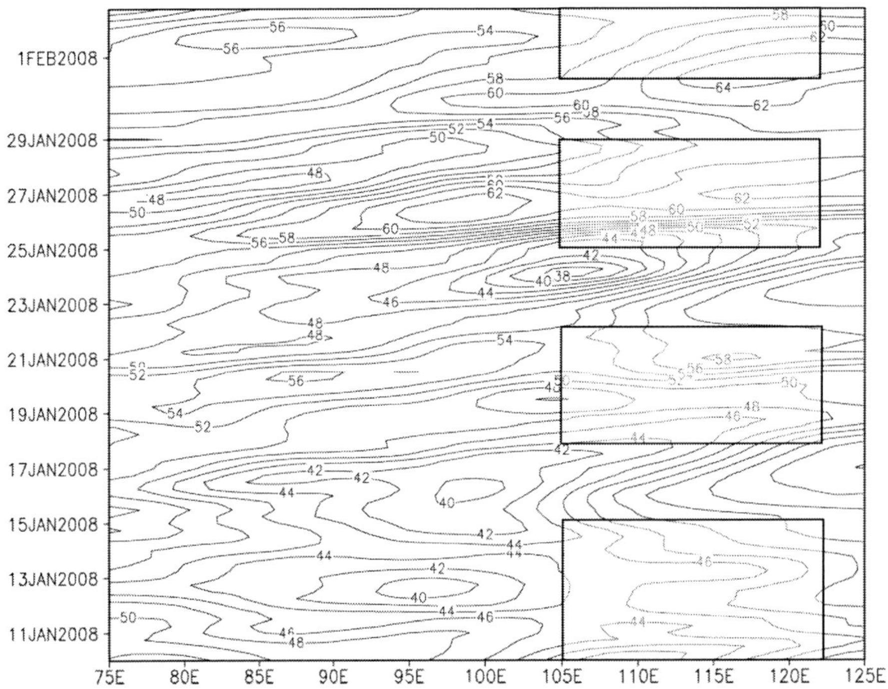

Figure 31. Meridional-time evolution of averaged (over 22.5°N–35°N) zonal wind at 200-hPa pressure level during 10 January–3 February 2008 (interval is 2 m/s). The shaded regions indicate the time and locations of surface snowstorms.

According to the distribution of wind fields at a 200-hPa level for the particular episode (Figure 27), the jet stream was located in a latitudinal belt of 22.5°N–35°N. In order to understand the relationship between the variational characteristics of the jet stream and surface snowfall for the entire storm period, we plotted a meridional-time evolution of zonal wind fields at the 200-hPa pressure level, averaged over the latitudinal belt (Figure 31). The shaded regions denote the locations (both in time and longitudes) of the four precipitation episodes. It can be seen that the jet stream intensified with time as the snowstorms developed from one episode to next. Significant precipitation occurred at the locations below the entrance region of the jet streak on the equatorial side, i.e., the heavy precipitation could be correlated to upper-level divergence in the right entrance region of the jet steak. As the jet intensified with time, the "pumping effect" due to the divergence field at the the right entrance region of the jet streak became dominant. In response, the storms began to organize into the classical configuration with a complete coupling of surface- and upper-level systems, and more precipitation followed. These features are consistent with the conceptual models of ageostrophic flows associated with the entrance and exit regions of coupled jets summarized in Figure 7-4 in KU.

Combining these analyses, an important thermodynamic feature stands out during these snowstorm episodes. The cold-dry air intrusion from the north seemed to play a central role in enhancing the baroclinic zone. This provided a favorable condition for a prolonged period of warm advection in the mid-troposphere far to the north along a quasi-stationary frontal boundary, thus initiating and maintaining the snowstorm. This cold-dry air intrusion formed a "cold wedge" (see Figure 23). Because this cold-air was wedged in at lower atmospheric levels, warm-moist air from south was forced upwards, providing a mechanism for cloud and precipitation to form. The formation of the "cold-wedge" is consistent with the pattern of the "north ridge-south trough" in the large-scale dynamic system (Figure 21a) as discussed in the previous section.

2.5. Conclusion

In this study, we assembled composite analyses as well as a case study for China's great snowstorms of 2008. Through these analyses, we obtained the following physical pictures of the large-scale dynamic and thermodynamic structures of the snowstorms (Figure 32).

From analyses of SLP and 500-hPa geopotential height fields, we found that a prominent blocking ridge existed over the northern Asian continent during these snowstorms. Meanwhile, all four snowstorm episodes presented a "north ridge-south trough" pressure distribution at 700 hPa, which led to a southward intrusion of cold air. The maintenance of a WPSH promoted the transport of warm-moist air into southeast China. The WPSH also played a crucial role in blocking the storm system from moving eastward out to sea. The confluence of the two air flows, the northern cold air and the warm, moist air from the south, provided a favorable condition for a prolonged period of warm advection in the mid-troposphere far to the north along a quasi-stationary frontal boundary. Further analyses of atmospheric vertical circulations indicated that there was strong vertical wind shear at low atmospheric levels. In the snowstorm region, northeasterly wind dominated below 850 hPa, while southwesterly wind prevailed above 850 hPa. Where the two airflows met, the warm-moist air was lifted up over the cold air. This process caused both a temperature and moisture inversion, and thus formed a "cold-air wedge" with a "warm-moist cover" sliding over a "cold-dry base." The upper-level jet stream was intensified, and was consistent with the intensifications of surface lows, in the sense that for the heavy precipitation events the right entrance region of the jet streak aligned with surface precipitation. All these in-phase dynamic and thermodynamic structures were responsible for sustaining and intensifying the storms, which evolved into one of the most extraordinary disastrous weather events in recent Chinese history.

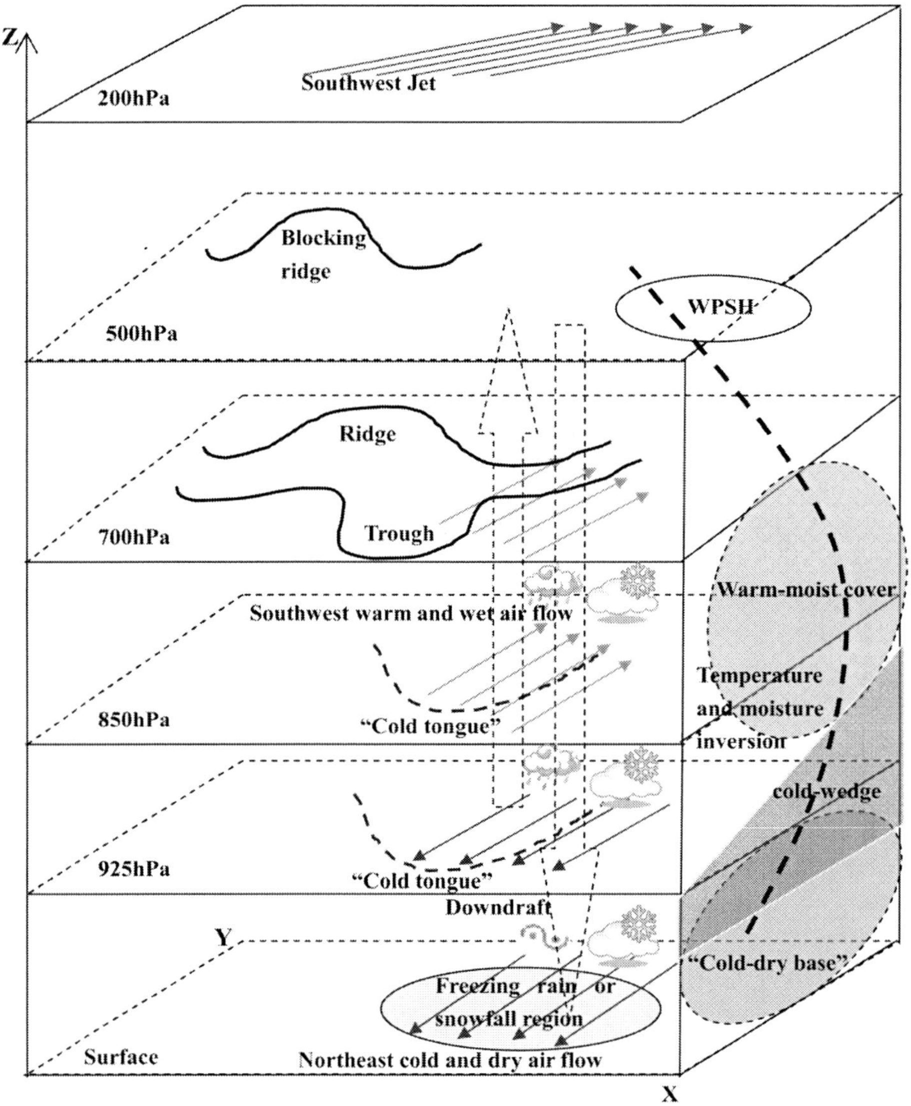

Figure 32. Conceptual diagram of dynamic and thermodynamic structures associated with China's great snowstorms of 2008.

ACKNOWLEDGMENTS

This work was Jointly supported by the International Sci.-Tech. Cooperative Project (2009DFB20540) and the social commonweal profession research program (GYHY201006009) funded by the Ministry of Science and Technology of the People's Republic of China, the Independent Research Project of LaSW (2008LASWZI04, 2009LASWZF02).

REFERENCES

Barbé, L. L., T. Lebel, and D. Tapsoba, 2002: Precipitation Variability in West Africa during the Years 1950–90. *J. Climate*, 15: 187–202.

Barlow, M., H. Cullen, and B. Lyon, 2002: Drought in central and southwest Asia: La Niña, the warm pool, and Indian Ocean precipitation. *J. Climate*, 15: 697–700.

Bordi, I., K. Fraedrich, and F. W. Gerstengarbe, 2004: Potential predictability of dry and wet periods: Sicily and Elbe - Basin (Germany). *Theor. Appl. Climatol.*, 77: 125– 138.

Bordi, I., K. Fraedrich, J. Jiang, and Coauthors, 2006: Spatio-temporal variability of dry and wet periods in eastern China. *Theor. Appl. Climatol.*, 79: 81 – 91.

Chagnon, F. J. F., and R. L. Bras, 2005: Contemporary climate change in the Amazon. *Geophys. Res. Lett.*, 32: L13703, doi:10.1029/ 2005GL022722.

Chen, L., X. Zhou, W. Li, and Coauthors, 2004: Characteristics of the climate change and its formation mechanism in China in last 80 years. *Acta Meteorologica Sinica* (in Chinese), 62 (5): 634–645.

Chen, W., and L. Kang, 2006: Linkage between the Arctic oscillation and winter climate over East Asia on the interannual timescale: roles of quasi-stationary planetary waves. *Chin. J. Atmos. Sci.* (in Chinese), 30 (5): 863 – 870.

Deng, Y., L. Cheng, and X. Zhang, 2000: Three-phase cloud explicity precipitation scheme and mesoscale numerical simulation for generated cause of the "96.1" snowstorm. *Plateau Meteor.* (in Chinese), 19 (4): 401-414.

Ding, Y., Q. Tu, and M. Wen, 1995: A statistical model for investigating climatic trend turning points. *Adv. Atmos. Sci.*, 12 (1): 47-56.

Fu, C., and Q. Wang Q, 1992: The definition and detection of the abrupt climatic change. *Chin. J. Atmos. Sci.* (in Chinese), 16 (4): 482–493.

Gong, J, and Z. Li, 2001: Low-level jet and heavy snow or snowstorm in Inner Mongolia. *Meteor. Mon.* (in Chinese), 27 (12): 3-7.

Gong, D., and S. Wang, 1999: Influence of ENSO on the precipitation over China and global land for the last century. *Chinese Science Bulletin*, 44 (3): 315 – 320.

Gong, D., and S. Wang, 2003: Influence of Arctic Oscillation on winter climate over China. *Acta Geographica Sinica*, 58 (4): 559-5680

Gong, D., S. Wang, and J. Zhu, 2004: Arctic Oscillation influence on daily temperature variance in winter over China. *Chinese Science Bulletin*, 49 (6): 637 - 642

Gong, D., J. Zhu, and S. Wang, 2002: The influence of Siberian high on Large-scale climate over continental Asia. *Plateau Meteor.* (in Chinese), 21 (1): 8-14

Hansen, J., R. Ruedy, M. Sato, and Coauthors, 2001: A closer look at United States and global surface temperature change. *J. Geophy. Res.*, 106: D20, 23947 – 23963.

Haylock, M. R., T. C. Peterson, L. M. Alves, and Coauthors, 2006: Trends in total and extreme South American precipitation in 1960 - 2000 and links with sea surface temperature. *J. Climate*, 19: 1490–1512.

He, X., Q. Li, Y. Ding, 2007: Numerical simulation of East Asian winter monsoon with ENSO episodes using a regional climate model. *Acta. Meteor. Sinica* (in Chinese), 65(1): 18-28

Huang, R., K. Wei, and J. Chen, 2007: The East Asian winter monsoon anomalies in the winters of 2005 and 2006 and their relations to the quasi-stationary planetary wave activity in the Northern Hemisphere. *Chinese J. Atmos. Sci.* (in Chinese), 31 (6): 1033 – 1048

Kalnay, E., M. Kanamitsu, R. Kistler, and Coauthors, 1996: The NCEP/NCAR 40-year reanalysis project. *Bull. Amer. Meteor. Soc.*, 77, 437–471.

Klein Tank, A. M. G., and G. P. Können, 2003: Trends in indices of daily temperature and precipitation extremes in Europe, 1946–1999. *J. Climate*, 16: 3665 – 3680.

Kocin, P. J., 1983: An analysis of the "Blizzard of 88." *Bull. Amer. Meteor. Soc.*, 64: 1258-1272.

Kocin, P. J., P. N. Schumacher, R. F. Morales, and L. W. Uccellini, 1995: Overview of the 12-14 March 1993 superstorm. *Bull. Amer. Meteor. Soc.*, 76: 165-182.

Kocin, P. J., and L. W. Uccellini, 2004: *Northeast Snow storms*. Meteor. Monogr., No. 54, *Amer. Meteor. Soc.*, 64, 818 pp.

Li, C., J. Zhu, and Z. Sun, 2002: The study of inter-decadal climate variation in China. *Climatic and Environmental Res.* (in Chinese), 7 (2): 209–19.

Li, Q., and Y. Ding, 2005: Multi-year simulations of East Asian winter monsoon by using regional climate model. *J. Appl. Meteor. Sci.*, 16 (Suppl): 30-40.

Li, Q., Y. Ding, W. Dong, 2007a: A numerical study on the winter monsoon and cold surge over East Asia. *Adv. Atmos. Sci.*, 24 (4): 664–678.

Li, Y., J. He, A. Jiang, 2007b: Circulation structure features of Western Pacific teleconnection pattern in winter and their relation with China's temperature and precipitation in winter. *Scientia Meteorologica Sincia*, 27 (2): 119-125.

Liebmann, B., C. S. Vera, L. M. V. Carvalho, and Coauthors, 2004: An observed trend in Central South American precipitation. *J. Climate*, 17: 4357 – 4367.

Liu, B., M. Xu, M. Henderson, and Coauthors, 2004: Taking China's Temperature: Daily Range, Warming Trends, and Regional Variations, 1955–2000. *J. Climate*, 17: 4453–4462.

Ma, Z., and Z. Fu, 2007: Evidences of drying trend in the global during the later half of 20th century and their relationship with large-scale climate background. *Science in China* (Series D), 50 (5): 776-788.

Mao, R, D. Gong, and Q. Fang, 2007: Influence of the East Asian jet stream on winter climate in China. *J. Appl. Meteor. Sci.*, 18 (2): 137-146.

Mitchell, T. D., and P. D. Jones, 2005: An improved method of constructing a database of monthly climate observations and associated high-resolution grids. *Int. J. Climatol.*, 25: 693–712.

National Resources Conservation Service, 2009: What is now water equivalent? United States Department of Agriculture. Availble at: http://www.or.nrcs.usda.gov/Snow/about/swe.html

New, M., D. Lister, M. Hulme, and Coauthors, 2002: A high-resolution data set of surface climate over global land areas. Climate Res., 21: 1–25.

New, M., M. Hulme, and P. D. Jones, 1999: Representing twentieth century space-time climate variability. Part 1: development of 1961-90 mean monthly terrestrial climatology. *J. Climate*, 12: 829–856.

New, M., M. Hulme, and P. D. Jones, 2000: Representing twentieth century space-time climate variability. Part 2: development of 1901-96 monthly grids of terrestrial surface climate. *J. Climate*, 13: 2217 – 2238.

Nicholson, N., 2004: The changing nature of Australian drought. *Climate Change*, 63: 323–336.

Nicholson, S. E., 2001: Climatic and environmental change in Africa during the last two centuries. *Climate Res.*, 17: 123–144.

Qian, W., H. Zhang, and Y. Zhu, 2001: Interannual and inter-decadal variability of East Asian Acas and theirs impact on temperature of China in winter season for the last century. *Adv. Atmos. Sci.*, 18 (4): 511–523.

Qian, W., and W. Zhang, 2007: Changes in cold wave events and warm winter in China during the last 46 years. *Chinese J. Atmos. Sci.* (in Chinese), 31 (7): 1266–1278.

Ren, G., J. Guo, M. Xu, and Coauthors, 2005: climate changes of China's mainland over the past half century. *Acta Meteorologica Sinica* (in Chinese), 63 (6): 942–955.

Shi, X., X. Xu, and L. Xie, 2007: Inter-decadal spatial-temporal change trend of East Asian winter monsoon in recent 40 years. *Chinese J. Atmos. Sci.* (in Chinese), 31 (4): 747-756.

Solow, A. R., 1987: Testing for climatic change: An application of the two-phase regression model. *J. Climate Appl. Meteor.*, 26: 1401–1405.

Steven, S. M., 2003a: Intra- to multidecadal climate variability over the continental United States: 1932 – 99. *J. Climate*, 16: 2215 – 2231.

Steven, S. M., 2003b: Multidecadal regime shifts in U.S. streamflow, precipitation, and temperature at the end of the twentieth century. *J. Climate*, 16: 3905 – 3916.

Tao, S., and Q. Zhang, 1998: Response of the Asian winter and summer monsoon to ENSO events. *Chinese J. Atmos. Sci.* (in Chinese), 22 (4): 399 – 407.

Tolika, K., and P. Maheras, 2005: Spatial and temporal characteristics of wet spells in Greece. *Theor. Appl. Climatol.*, 81: 71 – 85.

Tomé, A. R., and P. M. A. Miranda, 2004: Piecewise linear fitting and trend changing points of climate parameters. *Geophy. Res. Lett.*, 31 (2): doi: 10.1029/2003GL019100.

Trenberth, K. E., P. D. Jones, P. Ambenje, and Coauthors, 2007: Observations: Surface and Atmospheric Climate Change. In: Solomon S, Qin DH, Manning M, et al Eds. *Climate Change 2007: The Physical Science Basis. Contribution of Working Group I to the Fourth Assessment Report of the Intergovernmental Panel on Climate Change.* Cambridge, United Kingdom and New York, NY, USA: Cambridge University Press, 235–336.

Wang, H., 2001: The weakening of the Asian monsoon circulation after the end of 1970s. *Adv. Atmos. Sci.*, 18 (3): 376 -386.

Wang, W, and L. Chen, 2002: Numerical study on three-dimensional conditionally symmetric instability of "96.1" snowstorm event. *Plateau Meteor.* (in Chinese), 21 (3): 225-232.

Wang, W., and X. Xu, 1979: The heavy snow process in district Xilingele and the analysis of "77.10" snowstorm. *Acta Meteor. Sinica* (in Chinese), 37 (3): 80-86.

Wang, Z., and Y. Ding, 2006: Climate change of the cold wave frequency of China in the last 53 years and the possible reasons. *Chinese J Atmos Sci.* (in Chinese), 30 (6): 1068 – 1076.

Wang, Z., Y. Ding, J. He, and Coauthors, 2004: An updating analysis of the climate change in China in recent 50 years. *Acta Meteor. Sinica* (in Chinese), 62(2): 228–236.

Wu, B., and J. Wang, 2002: Possible impacts of winter arctic oscillation on Siberian High, the East Asian winter monsoon and sea-ice extent. *Adv. Atmos. Sci.*, 19 (2): 297 – 320.

Xu, J., Q. Zhu, and T. Zhou, 1999: Sudden and periodic change of East Asian winter monsoon in the past century. *J. Appl. Meteor. Sci.* (in Chinese), 10 (1): 1-8.

Yang, L., C. Miao, and S. Shou, 2006: Numerical simulation on a snowstorm in Yangtze-Huaihe River Basin in spring 2003. *J. Nanjing Insti. Meteor.* (in Chinese), 29 (3): 379-384.

Yang, S., K. M. Lau, and K. Kim, 2002: Variations of the East Asian jet stream and Asian-Pacific-American winter climate anomalies. *J. Climate*, 15, 306 – 324.

Yu, S., X. Lin, and X. Xu, 2003: The Climatic change in northwest China in recent 50 years. *Climatic and Environmental Res.* (in Chinese), 8 (1): 9 – 18.

Zhang, X., and L. Cheng, 2000a: Dynamic diagnose of the genesis and development for mesoscale shear line during "96.1" snowstorm I: Diagnoses of vorticity and vorticity variability. *Plateau Meteor.* (in Chinese), 19 (3): 285-294.

Zhang, X., and L. Cheng, 2000b: Dynamic diagnose of the genesis and development for mesoscale shear line during "96.1" snowstorm II: Diagnoses of divergence and its variability. *Plateau Meteor.* (in Chinese), 19 (4): 459-466.

Zhou, L., and R. Huang, 2003: Research on the characteristics of inter-decadal variability of summer climate in China and its possible cause. *Climatic and Environmental Res.* (in Chinese), 8 (3): 274–290.

Zhu, A., and S. Shou, 1994: Diagnosis of frontgenetical secondary circulation in a winter snowstorm event. *J. Nanjing Insti. Meteor.* (in Chinese), 17 (2): 183-187.

Zou, X., P. Zhai, and Q. Zhang, 2005: Variations in droughts over China: 1951–2003. *Geophys. Res. Lett.*, 32: L04707, doi: 10.1029/2004GL021853.

In: Precipitation: Prediction, Formation, Environmental Impact ISBN: 978-1-62100-447-9
Editors: Henry Dohring and Jeremy Dixon

Chapter 3

STOCHASTIC MODELING AND SIMULATION OF DAILY PRECIPITATION FOR WEATHER GENERATION

W. W. Ng[1,2], U. S. Panu[2,1] and T. Sharma[2]
[1]University of Manitoba, Winnipeg, Canada
[2]Lakehead University, Thunder Bay, Canada

ABSTRACT

Weather generation is critical when the historical data is not comprehensive for the required analysis. Therefore, the purpose of weather generation is to generate synthetic sequences of weather variables (e.g., precipitation) that are statistically indistinguishable from the observed characteristics of the historical records. These generated sequences provide a set of alternative realizations for use in the design, operation, management, and policy decisions related to water resources and hydrological systems. The significance of weather generation, for example, can easily be appreciated by examining different research areas, such as modeling of weather, analysis of climate-sensitive systems, and impact assessment of scenarios on water balance in the wake of impending climate change.

Weather generation normally implies simulation of precipitation data, which serves as an important input to hydrological systems and is of importance in local rainfall-runoff studies [Yoo and Ha, 2002]. To be able to capture the statistical characteristics of the precipitation data into the model, (1) the unique characteristics of the data set, (2) the temporal dependency in a precipitation series, and (3) the spatial dependency of precipitation amount among stations have to be modeled properly.

Due to the unique characteristics of the precipitation data set, modelling of daily precipitation data commonly faces with several challenges, such as: (i) the difficulty of accommodating the occurrence of sustained dry and wet periods; (ii) the prevalence of missing and trace records causing unspecified uncertainty in analyses; (iii) the familiar probability distributions incapability of modelling the precipitation time series, as such series tend to be discontinuous comprised zeros (occurring intermittently) and non-negative numbers.

In respect to the modeling of temporal dependency, precipitation sequences often have been modelled using a short-memory model, but results have not always been satisfactory. Some researchers have advocated the inclusion of higher order memory components for improving the simulation results in terms of long-term dependencies [Buishand, 1978; Rolda'n and Woolhiser, 1982; Semenov, et al., 1998] but such an approach may result over parameterized the model.

Likewise, the modelling on a spatial basis faces problems because of the occurrence of non-Normal probability distributions of precipitation data. If precipitation data were normally distributed, it would be relatively easy to develop a spatial linkage among the adjoining precipitation stations in the form of multivariate Normal distribution. However, precipitation data have a lower bound of zero and are skewed distributed that makes it difficult to establish a multivariate relationship among the precipitation data of different stations.

Precipitation series have been conventionally regarded as a stationary process. Pragmatically, one cannot ignore the influence of global warming on precipitation, particularly at the present juncture when rising global temperatures are causing alarms and are impacting atmospheric processes [IPCC, 2007]. Thus, the influence of climate change to the modeling of precipitation data is drawing more attention for the last few decades. In view of the above challenges, this chapter aims to present the concepts and methods that can preserve the statistical characteristics of precipitation for the weather generation.

1. Generation of Daily Precipitation Sequences at a Single Station

A point generation process of daily precipitation should preserve historical features such as temporal dependence in successive occurrences of wet and dry days, and the underlying statistical properties of the probability distribution of the amounts (i.e., mean, variance, and skewness). Therefore, modeling of daily precipitation at a single station may normally consist of two components, namely occurrences and amounts. These two components are modeled separately [Rolda'n and Woolhiser, 1982; Woolhiser and Rolda'n, 1982] and then coupled to provide a synthetic sequence of data [Richardson and Wright, 1984; Cowpertwait, 1991].

Modeling of Precipitation Occurrences

Conceptually, the generation of precipitation occurrences relies on the assumption that there is an existence of short-term temporal-dependency between the neighbouring states while implying the existence of long-term independency of states in precipitation records. For example, the dry/wet-spell length of an alternating renewal process (ARP) is formulated based on the assumption of persistency (short-term temporal-dependency) within a time series [Racsko et al., 1991]. However, the occurrences of dry and wet-spell lengths are independent (long-term temporal-independency) of each other and the independency among dry/wet-spell lengths can be observed from the random sampling alternatively between wet and dry spell distribution. Similarly, the conditional transition probabilities of a Markov chain model preserve the short-term temporal-dependency (chain-dependency) between the neighbouring states in a sequence. However, the future state is conditionally independent of the past states (but can be dependent on the current state) beyond those used specifically for the conditioning. Thus the basic structure of a generation model consists of two mechanisms: [1] a mechanism that can preserve the short-term temporal-dependency through grouping of the nearest sequential states into meaningful segments (e.g., spell length or transition states), and [2] a mechanism that assumes the independency among the meaningful segments through sampling from probability distributions.

The ARP and Markov chain models, among others, are extensively used because of their simplicity and parsimony in modeling. The generation of weather dry/wet states using the ARP model relies on sampling dry/wet-spell lengths from their corresponding probability distributions of dry and wet states, which are selected based on the goodness-of-fit of the observed dry/wet spell lengths [Rolda'n and Woolhiser, 1982; Semenov and Barrow, 1997]. The selected distribution functions could be an empirical, truncated negative binomial or geometric distribution [Buishand, 1978; Rolda'n and Woolhiser, 1982]. On the other hand, the Markov chain models display a wide range of applications [Bailey, 1964; Smith, 1987; Hay et al. 1991; Salas, 1993] and have been used in the development of a variety of models, e.g., Markov renewal model [Foufoula-Georgiou and Lettenmaier, 1987], Hidden Markov model [Salas and Boes, 1980; Zucchini, and Guttorp, 1991; Thyer and Kuczera, 2000; Thyer and Kuczera, 2003a; Thyer and Kuczera, 2003b], and non-homogenous Hidden Markov model [Hughes and Guttorp, 1994a; Hughes et al. 1999; Bellone et al., 2000; Lambert et al., 2003, Rasmussen and Akintug, 2004]. It is in this vein that these models have been widely used as components of many existing stochastic weather generators (e.g., the ARP used in the LARS-WG [Racsko et al., 1991; Semenov and Barrow, 1997, Semenov et al., 1998] and the Markov chain used in the WGEN [Richardson and Wright, 1984], SIMMETEO [Geng, et. al., 1986], WeatherMan [Pickering et al., 1994], and WXGEN [Hayhoe and Stewart, 1996]).

Early applications of the Markov chain models proved to be simple but represented a reasonable simulating mechanism for the generation of precipitation occurrences [Gabriel and Neumann, 1962]. A first-order Markov chain model was found to be better than a Bernoulli model for describing the occurrences of dry/wet days [Smith and Schreiber, 1973]. However, the Markov chain models have also been found to be inappropriate in modeling the clustering tendencies that exist in daily precipitation sequences [Foufoula-Georgiou and Lettenmaier, 1987]. These models are said to possess a limited memory of rare events that could fail to accurately simulate the long dry series at certain locations [Racsko et al., 1991]. Moreover, the use of Markov chain models in meteorological data has been shown to underestimate the observed relative frequency of long dry spell lengths [Racsko et al., 1991] because the probability of occurrence of long dry or wet-spell length decreases exponentially in the Markov chain models [Feller, 1970]. On the other hand, the performance of ARP models were considered to be comparable to the Markov chain models [Racsko et al., 1991; Semenov et al., 1998]. However, other studies have reported the superiority of the Markov chain models based on the minimum Akaike information criterion (AIC) [Rolda'n and Woolhiser, 1982]. Because the characteristics of meteorological data are functions of the location and season, the results of different studies are varied. In addition, the results obtained based on different guiding principles (e.g., matching the observed dry/wet-spell probability distributions or AIC to those of the generated data) may reveal unique advantages of different models for specific aspects. Additional comparisons between the ARP models using Geometric distribution and the Markov chain can also be found elsewhere [Buishand, 1978; Rolda'n and Woolhiser, 1982].

Despite the popularity of the ARP and Markov chain models, they are still considered inflexible and non-robust in capturing the variation in the dependence structure of precipitation sequences. Existing temporal-dependent models for the generation of precipitation occurrences are restricted either by the fixed-length memory (e.g., the order of a Markov chain model), or by the reining states in segments (e.g., persistency of homogenous states within dry/wet-spell lengths of an ARP) [Ng et al., 2010]. The modeling of variable segmented lengths can be accomplished through a hybrid-order Markov model [Stern and Coe, 1984; Wilks and Wilby, 1999b]. The hybrid-order model is a higher-order Markov model that reduces the number of parameters through generalization of the possible segment-types. The hybrid-order models retain first-order dependence for wet spells, but allow higher-order dependence among dry spells. Such dependence is useful when long memory of dry spells is desired. At times, however, the long wet spells sequence could be of interest as excessive wet weather spell may prevent ripening of grains and/or harvesting of agricultural crops.

Recently, an innovative Dictionary approach [Pachter et al., 1999; Bussemaker et al., 2000] has been developed in the field of genome pattern recognition for the identification of frequently occurring genome segments in DNA sequences. The genome segments delineate the biologically meaningful "words" (i.e., segments with a specific patterns in a series of discrete states) that can be jointly modeled with variable lengths and states. Such flexibility would provide a unique advantage over the traditional stochastic models for the generation of precipitation occurrences. Although the Dictionary approach was originally adopted in the field of genome pattern reorganization, the legitimacy of using such an approach in modeling daily precipitation can be supported based on that there is a similarity in structure of the Dictionary approach to the existing precipitation synthesis methods (e.g., Markov chain and alternating renewal process). Existing precipitation synthesis methods commonly consist of a mechanism to preserve the short-term dependencies of historical data (e.g., transition probability of Markov chain) and another mechanism to allow the random process to be involved in the generation of the long-term sequence (e.g., drawing samples from dry or wet states). The collective function of these two mechanisms is to preserve the physical characteristics of daily precipitations that normally possess a stronger short-term temporal-dependency while still retaining a weaker long-term temporal-dependency. In other words, to a certain extent the methods that possess these two mechanisms, can explicitly preserve the physical characteristics of daily precipitations (i.e., the process) and since the Dictionary approach does possess these two mechanisms in its structure, therefore it would implicitly preserve these characteristics [Ng and Panu, 2010].

Modeling of Precipitation Amounts

After having modeled the state (i.e., wet or dry day), the next step is to model the amounts of precipitation on the wet days using a suitable probability distribution function. A probability distribution is selected based on the goodness-of-fit between empirical (based on the observed amounts) and its corresponding theoretical counterpart. Once a suitable probability distribution has been estimated the wet state precipitation amounts can be generated through random sampling from the distribution. The two-parameter gamma distribution [Katz, 1977; Buishand, 1978; Stern and Coe, 1984] and the three-parameter mixed exponential distribution [Smith and Schreiber, 1974; Woolhiser and Rolda'n, 1982; Wilks and Wilby, 1999b] have been the common choices for modeling of the precipitation amounts. Several different distribution models have been proposed and compared [Woolhiser and Rolda'n, 1982] for synthesizing the precipitation amounts.

Analyses relying solely on models designed for single sites are inadequate to meet the data requirements of large water resource systems (e.g., drought risk assessments, operation and management of large reservoirs, etc.) and found to be weak in terms of preserving the attribute of long term persistence [Thyer and Kuczera, 2000]. The precipitation at multiple sites can be generated site by site by adopting the single site based algorithms. The averaging of precipitations at multiple sites provides an estimate of the spatial precipitation. In such a procedure, however, the spatial dependency among stations is ignored, which raises a question as to the ability of generated sequences representing the spatial characteristics of precipitation, including inter-station correlation and non-homogeneity.

2. Generation of Daily Precipitation Sequences at Multiple Stations

Precipitation modeling at multiple sites is a daunting task due to the complexities of spatial dependencies and non-homogeneities, which is probably the reason for the limited literature available in this area [Srikanthan and McMahon, 2001]. A lack of literature and the need for spatially distributed data have recently spurred the interest for weather generation at multiple sites [Wilks, 1999a]. The development of a spatially dependent weather generating model is basically intended to achieve two goals: (1) to provide a means of generating synthetic precipitation time series reproducing the spatial statistical characteristics of the observed data at multiple sites, and (2) to provide a means of extending the precipitation time series at ungauged locations [Semenov and Barrow, 1997; Wilks, 2008]. Many of the weather generation models (e.g., NSRP, weather generator (WGEN), hidden Markov model, *k*-nearest neighbour) designed for multiple sites [Cowpertwait et al., 2002; Wilks, 1998; Thyer and Kuczera, 2003a; Buishand and Brandsma, 2001] are based on the methods applicable to a single site [Bras and Rodriguez-Iturbe, 1976; Smith and Karr, 1985; Cox and Isham, 1988; Cowpertwait, 1995; Richardson and Wright, 1984; Hughes et al., 1999; Rajagopalan and Lall, 1999]. However, models designed for a single site tend to be inadequate in capturing the spatial dependence, other alternative approaches are being pursued to address the spatial dependency issue. Various weather generation methods are broadly categorized as either parametric or non-parametric. These methods are basically an extension of approaches based on the univariate distribution into a multivariate domain to account for dependency among sites.

Parametric Approaches for Normally Distributed Observations

The multivariate truncated Normal distribution model is an extensively used and promising parametric approache for modeling precipitation data at multiple sites [Scheuer and Stoller, 1962; Bell, 1987; Bardossy, 1992; Hutchinson and Richardson, 1993; Hutchinson, 1995; Sansom and Thomson, 1998; Jothityangkoon et al., 2000; Stehlik and Bardossy, 2002; Sanso and Guenni, 2004]. An important feature of the truncated Normal distribution approach is that the observed daily precipitation data containing zero and positive values can be transformed into a normally distributed data set, which is well amenable to statistical analyses. A model in the form of the truncated Normal distribution [Bardossy, 1992] has been devised by transferring precipitation values equal to zero to data values in the negative range such that a Normal distribution can be constructed. The maximum-likelihood method [Cohen, 1959] is commonly used for the estimation of parameters of a univariate truncated Normal distribution. Formulating a space-time models invokes concepts of time series analysis (i.e., multivariate autoregressive moving average model).

Modeling with the unique characteristics of daily precipitation may sometimes be challenging. An ordinary precipitation data set contains several problematic records (i.e., zero-, extreme-, trace-, and missing values) that can limit the usefulness of a Normal distribution. The zero records are abundant and not uncommon in dry regions as well as in dry seasons. The extreme values and outliers are easily discernable in daily precipitation records. Recently, researches [Panu et al., 2003; Ng et al, 2007] have demonstrated the utility of various outlier detection methods in a hydrologic context. The existence of trace records is mainly due to the limitations of the precipitation measuring devices. The trace precipitation could be a measurable amount in the range greater than zero and less than 0.25 mm. Some researchers simply treat trace records as zero because of their insignificant magnitude while others round them to an average value of (0.20 mm/2) or 0.10 mm [Yang et al., 1998]. Their exact amounts are uncertain and the impact may sometimes be significant [Benson, 1982]. Likewise, missing data are also common in records for varied reasons and pose problems in modeling and simulation studies of precipitation time series [Ng et al., 2009]. In short, records with zero, extreme, trace, and missing values can seriously affect the fitting of statistical models leading to enhanced uncertainty in model outputs.

The transformation of the problematic records can be handled by several procedures in association with the truncated Normal distribution approach. Firstly, the above-zero records are normalized by a Box-Cox power transformation and the non-normality impact of the extreme data can be minimized. Secondly, the parameters of a Normal distribution are estimated using incomplete data (i.e., above-zero records). The parameters of the Normal distribution thus estimated are based on the concept of maximum likelihood derived from the univarite truncated Normal distribution [Cohen, 1959]. The parameters of multivariate Normal distribution can be estimated based on minimizing the differences between the empirical distributions (i.e., histogram) and the theoretical distribution functions (i.e., probability and cumulative density functions) in association with the truncated Normal distribution. Trace and missing records are substituted based on the conditional information (e.g., the spatial correlation) of the multivariate Normal distribution. As all the problematic records are substituted, the parameters of the complete transformed data set can be estimated. Generation of the daily precipitation data at multiple sites can be achieved simply by randomly drawing samples from the multivariate distribution model using the estimated parameters. The proposed method can replace all of the trace and missing records from the data set and a complete normally distributed data set can be obtained.

Although the truncated Normal distribution model appears attractive, the estimation of the parameters of the distribution is not trivial. The estimation procedures become more complicated when the parameters of a multivariate truncated distribution must be estimated. Moreover, many of the estimation procedures are designed for the estimation of parameters of a specific distribution function (e.g., Normal function) which is considerably restrictive.

Therefore, a generalized and simple procedure for estimation of parameters of the truncated distribution is desirable.

The modeling of spatial dependency is not limited to normally distributed data. However, the methods for estimation of parameters of non-Normal multivariate probability distributions are far less developed. That is one reason why the multivariate truncated Normal distribution is often used for modeling the precipitation at multiple sites.

Parametric Approaches for Non-Normally Distributed Observations

The statistical methods for multivariate Normal distributed data in which the dependency is represented in a linear fashion (i.e. correlation coefficient) are reasonably well developed. Studies on the representation of non-linear dependency and modeling of non-normally distributed data are relatively rare.

One limitation of the conventional joint distribution functions is their inability to preserve the variable association (i.e., dependency) in the entire frequency range. To elucidate the point, one can consider the example of precipitation data at two stations which can be fitted by a bi-variate Normal probability function with 5 parameters (i.e., two means, two variances, and a correlation value) wherein spatial dependency is reflected by the correlation value. It is generally known that storms with high precipitation may cover larger areas than those with low precipitation. Therefore, intuitively the inter-station correlation in the low range of precipitation is likely to be different from that of the high range of precipitation. The issue of variable levels of dependency is difficult to address with the conventional multivariate Normal distribution modeling approach.

Recently, a method known by the name of Copula has been developed, which appears promising in order to model the spatial dependencies in the form of joint probability distributions [Favre et al., 2004]. Mathematically, the Copula [Sklar, 1959] connects univariate marginal distributions to a joint distribution using the Copula function with a non-linear dependence structure. The Archimedean function is selected for the formation of the Copula model because of its simplicity and popularity in the field of Copula studies related to the modeling of precipitation data [Nelsen, 1999]. This unique dependency feature along with its flexibility in modeling the relationship between the marginal and joint distributions gives an edge to the Copula as a spatial-dependent approach for the generation of precipitation.

Many techniques related to univariate Copula have been developed. However, studies related to multivariate Copula are somewhat limited because of the complex structure of mathematical functions involved in the analysis [Nelsen, 1999]. As a result, a very limited number of multivariate Copula formulations are available that can be reformulated to represent daily precipitation at multiple sites. A common concern with the use of the Copula approach is that it requires a priori selection of probability distribution functions such that determining the corresponding values of parameters could be time intensive to estimate and may involve subjectivity.

Non-Parametric Approaches for Non-Normally Distributed Observations

A relatively effective and objective approach for the modeling of spatial-dependent precipitation is through using non-parametric approaches. Several non-parametric weather generation methods based on the spatial dependency have been developed during the last few decades. Some studies use kernel density estimators and *k*-nearest neighbour re-sampling to represent the probability distributions of wet day precipitation amounts at multiple sites [Lall et al., 1996b; Harrold et al., 2005]. Other studies [Mehrotra and Sharma, 2007] apply a semi-parametric approach to investigate the variability of the low-frequency of precipitation at multiple sites. Extrapolation and interpolation are also two simple non-parametric approaches used for the generation of spatially dependent precipitation patterns. For example, Semenov and Brooks [1999] adopted the spatial interpolation approach with the Long Ashton Research Station Weather Generator (LARS-WG) model to generate weather patterns in Great Britain.

Among all non-parametric approaches, the k-nearest neighbour re-sampling [Young, 1994; Sharma and Lall, 1999; Rajagopalan and Lall, 1999; Buishand and Brandsma, 2001; Harrold el al., 2005] is a relatively popular technique used in stochastic weather generation. The k-nearest neighbour establishes a probability density environment (i.e., state space) to preserve the dependent relationship between predictor and predictant. The precipitation amounts at current stage (i.e., predictant) can be estimated based on the conditional information of the precipitation amount at the previous stage (i.e., predictor) through re-sampling from the k nearest observations in state space. The estimation solely relies on k number of observations that may over emphasize the impact of a small number of observations (e.g., the k values corresponding to the k-nearest neighbours). This is particularly important in the case of daily precipitation data riddled with outliers. In cases involving extreme data, it is desirable to develop a generalized non-parametric approach for the adequate modeling of daily precipitation records at multiple sites.

The parametric approaches for normally distributed data as well as non-parametric approaches for non-normally distributed data have been developed based on different concepts and formulation structures. Any one of these approaches can be combined with the temporal-dependent approach for the generation of a spatio-temporal precipitation at multiple sites. The generated data is expected to preserve statistical features of the historically observed precipitation records.

Noticeable changes in regional precipitation patterns over the past few decades due to climatic change have been the focus of attention of many researchers. Many of the existing weather generation models are capable of incorporating the impact of climatic change on precipitation. One such weather generation model will be utilized for downscaling of appropriate climatic variables as described in the ensuing sections.

3. Precipitation Model with Exogenous Variables - Statistical Downscaling

Precipitation modeling with exogenous variables refers to the generation of local precipitation data involving external variables, e.g., atmospheric temperature, radiation, pressure, and remote sensing data. Many recent studies related to weather generation involve

variables obtained from physically-based numerical climatic models (e.g., global climate models or regional climatic models) [Fowler et al., 2000]. The output from the climate models can be integrated into precipitation models to enhance the accuracy of the generated data in describing the future impact of climatic change at the local scale.
A special case is that of downscaling output from numerical climate models (e.g., global climate models (GCMs)) in which precipitation data forms the basis of downscaling to the desired local scale. The outputs (e.g., simulated monthly precipitation) from the climate models are typically reported with a low spatial resolution (e.g., 250 * 250 km for a GCM), which is inadequate for use in studies related to local hydrologic and water resource systems. Therefore, statistical downscaling is needed to improve the spatial resolution of output from climatic models.

Weather generation methods can be based on either the observed records obtained from local monitoring sites or the empirically generated observations in association with external variables. Many statistical downscaling approaches are similar to the traditional weather generating models, except that their parameters are influenced by external variables obtained from climate models [Wilks, 1992]. Downscaling of output of climate models to local precipitation has been studied [Hay et al., 1991; Wilson and Skyllingstad, 1992; Hughes and Lettenmaier, 1993; Hughes and Guttorp, 1994a; Hughes and Guttorp, 1994b; Hughes et al., 1999]; recent advances in climatic downscaling and the application of appropriate hydrological models for use in impact studies are also reported by Schulze [1997] and Rummukainen [1997]. Downscaling approaches can usually be divided into two categories: dynamic and statistical. The regional climate models (RCM) and the limited-area models (LAM) are examples of dynamic downscaling in which the climatic data are modeled into a finer grid scale (for example 20 * 20 km). Although dynamic downscaling has been reported to be successful in several studies [Giorgi and Bates, 1989; Giorgi, 1990; Giorgi and Mearns, 1991; Hostetler and Giorgi, 1992; Jones et al., 1995], it is computationally intensive. On the other hand, statistical downscaling has proved to be a good alternative for transformation of climatic variables [Hulme et al., 1993; Airey and Hulme, 1995]. Statistical downscaling models are well described in the literatures [Kim et al., 1984; Wigley et al., 1990; Grotch and MacCracken, 1991; Bardossy, 1992; von Storch et al., 1993].

Statistical downscaling techniques [Wilby and Wigley, 1997b] have been categorized as: (a) pattern classification methods, (b) regression methods, and (c) stochastic weather generators. The comparisons of these three schemes can be found in recent studies [Xu, 1999; Giorgi et al., 2001; Wilby et al., 2004].

Pattern Classification Method

In downscaling studies, pattern classification has been used to group days with similar weather characteristics into finite synoptic patterns. Many studies are related to regional precipitation associated with synoptic weather patterns [Lamb, 1972; Hay et al., 1991; Bardossy, 1992; Hughes et al., 1993; Jones et al., 1993; Bardossy et al., 1994; Bardossy et al., 1995; Wilby, 1995; Wilby, 1997]. Other studies are related to long-term temperature fluctuations [Sowden and Parker, 1981; Moses et al., 1987] and river flows [Duckstein and Bardossy, 1993; Wilby, 1993] with the synoptic weather patterns.

Some concerns associated with this scheme are worth noting. (1) Observations may not necessarily contain significant and distinguishable features that allow for reasonable clustering. (2) The procedure may involve some subjectivity in the selection of features, predictors, and number of classes that need to be pre-defined based on expert judgments [Wilby, 1994]. (3) The classification approach cannot handle circulation patterns, which do not fit well with the current classification. (4) The classification approach may not work for future climate [Wilby et al., 2004]. (5) The classification-based methods do not seem very effective in reproducing the persistent characteristics of wet-dry spells at a single site [Wilby, 1994].

Regression Method

In downscaling studies, relationships can be established between the output from numerical climatic models (independent variables) and the local observations (dependent variables). Multiple regression [Murphy, 1999], canonical correlation analysis [von Storch et al., 1993], and artificial neural networks analysis [Hewitson and Crane, 1992; Hewitson and Crane, 1996] are regression methods commonly used in climatic downscaling. The predictors and predictands in regression equations can be of the same type of variables [Wilks, 1989; Brown et al., 1995; Carbone and Bramante, 1995; Perica and Foufoula-Georgiou, 1996] or a different type of variables [Wigley et al., 1990; Zorita et al., 1992; von Storch et al., 1993; Kaas, 1993; Kaas, 1994; Johannesson et al., 1995; Conway et al., 1996]. A lucid discussion of the significance, problems, and issues associated with the regression approach is given elsewhere [Crane and Hewitson, 1998; von Storch, 1999].

The regression approach offers the advantage of being relatively simple to apply and also do not involve many operations. On the other hand, one of the disadvantages of regression methods is that they require many conditions and assumptions to be satisfied. For example, the methods require long observation series for validation and implicitly assume the normality of data, and therefore may be applicable to only selected sites and/or certain variables and seasons [Xu, 1999; Wilby et al., 2004]. The problem of under-prediction of variance and poor representation of extreme events often occurs when regression approaches are adopted [Burger, 2002].

Stochastic Weather Generators

The influence of climate change must be incorporated in the analysis. One possibility is to involve the output from regional climate models (RCMs). The precipitation data from RCMs are generated using complex mathematical algorithms involving physics of ocean-atmospheric processes on a global scale. For practical use, it is necessary to downscale relevant data from the global scale to the local scale and to couple them with models designed for preserving temporal and spatial dependencies in generated data sequences.

Stochastic weather generators are designed to preserve and reproduce the statistical characteristics (e.g., mean, variance, skewness, serial correlation) of local weather variables [Wilks, 1999b]. Examples of stochastic downscaling models include WGEN [Richardson and Wright, 1984; Wilks, 1999a] and LARS-WG [Racsko et al., 1991; Semenov and Barrow,

1997]. Two kinds of application can be found in literature related to stochastic weather generation: (1) generation of present climate [Richardson, 1981; Foufoula-Georgiou and Lettenmaier, 1987; Smith, 1987; Zucchini and Guttorp, 1991] and (2) generation of future climate for climatic change and impact studies [Wilks, 1992; Gregory et al., 1993; Katz, 1996; Mearns et al., 1996; Semenov and Barrow, 1997].

The advantage of stochastic weather generators is that the methods can generate long series of weather variables (e.g., precipitation series) with statistical features that are consistent with those of historical observations. However, stochastic weather generators may be unreliable when predicting changes in the mean level of precipitation. This is in particular critical to studies related to hydrology [Mearns et al., 1996]. The adjustment of parameters subject to the future climatic change could be arbitrary [Wilby et al., 2004] or may result in unexpected outcomes [Wilks, 1992].

In general, the limitations of statistical downscaling in local weather generation have rarely been considered [Giorgi and Mearns, 1991; Wilby and Wigley, 1997b]. Several limitations inherent in using statistical downscaling have been identified in the literature. For example, statistical downscaling methods are sensitive to the selection of the season, choice of the data standardization technique, the functional form, and the choice of predictor variables. The assumption of the stationary nature of the predictor-predictand relationship has also been criticized [Wilby and Wigley, 1997]. The preservation of low-frequency features in synthetic data is questionable [Katz, 1996]. Using global climate model (GCM) to predict changes in mean precipitation is known to be unreliable and this problem has not yet been resolved [Mearns et al., 1996].

5. Summary of Approaches and Models in Daily Precipitation Generation

Many studies in the domain of weather generation have focused on the generation of precipitation because of its importance. Precipitation data generation has been extensively addressed in the hydrologic and meteorological literature. The characteristics of precipitation data include the high frequency of zeros and the existence of missing/trace/extreme records, forming an intermittent data series that complicates the statistical modeling. The daily precipitation has two components viz. occurrence of dry day or wet day in a sequence and the amount (depth, mm of precipitation on the wet day). The occurrence component is usually modeled using notions of the Markov chains and/or alternating renewal processes. The amount component is commonly modeled through one, two or three parameter non-normal probability distribution functions. The spatial dependency is an important element and is basically modeled through the use of both parametric and non-parametric approaches. These models are implemented in an implicit manner that results in generated sequences of daily precipitation series at multiple sites. In addition to temporal and spatial dependencies in the precipitation data, there may be another dependency involving exogenous variables, typically output from a GCM, which needs to be downscaled for applications in hydrological analyses.

REFERENCES

Airey, M., & Hulme, M. (1995). Evaluating climate model simulations of precipitation: methods, problems and performance. *Progress in Physical Geography*, 19, 427-448.

Bailey, N. T. J. (1964). The elements of stochastic processes, John Wiley, New York.

Bardossy, A. (1992). Space-time model for daily rainfall using atmospheric circulation patterns. *Water Resources Research*, 28, 1247-1259.

Bardossy, A., Matyasovszky, I., & Bogardi, I. (1994). Knowledge based classification of circulation patterns for stochastic precipitation modelling. In K. W. Hipel, A.W. McLeod, U.S. Panu, and V. P. Singh, editors, *Stochastic and statistical methods in hydrology and environmental engineering,* volume 3, Kluwer, Dordrecht, Netherlands.

Bardossy, A., Duckstein, L., & Bogardi, I. (1995). Fuzzy rule-based classification of atmospheric circulation patterns. *International Journal of Climatology*, 15, 1087-1097.

Bell, T. L. (1987). A space-time stochastic model of rainfall for satellite remote-sensing studies. *Journal of geophysical research*, 92, 9631-9643.

Bellone, E., Hughes, J. P., & Guttorp P. (2000). A hidden Markov model for downscaling synoptic atmospheric patterns to precipitation amounts. Climate Research, 15, 1-12.

Benson, C. S. (1982). Reassessment of winter precipitation on Alaska Arctic slope and measurement on the flux of wind blown snow, Geophysical Institute, University of Alaska, Fairbanks.

Bowden, G. J., Maier, H. R., & Dandy, G. C. (2005). Input determination for neural network models in water resources applications. part 2. case study: forecasting salinity in a river. *Journal of Hydrology*, 301, 93-107.

Bras, R., & Rodriguez-Iturbe, I. (1976). Rainfall generation: a nonstationary time varying multidimensional model. *Water Resources Research*, 12, 450-456.

Brown, B. G., Mearns, L. O., & McDaniel, L. (1995). *Climatological characteristics of spatial scaling of hourly precipitation, paper presented at the Ninth Conference on Applied Climatology,* American Meteorological Society, 1520 January, Dallas.

Buishand, T. A. (1978). Some remarks on the use of daily rainfall models. *Journal of Hydrology*, 36, 295-308.

Buishand, T. A., & Brandsma, T. (2001). Multisite simulation of daily precipitation and temperature in the Rhine basin by nearest-neighbor resampling. *Water Resources Research*, 37, 2761-2776.

Burger, G. (2002). Selected precipitation scenarios across Europe. *Journal of Hydrology*, 262, 99-110.

Bussemaker, H. J., Li, H., & Siggia, E. D. (2000). Building a dictionary for genomes: Identification of presumptive regulatory sites by statistical analysis. *Proceedings of the National Academy of Sciences of the United States of America*, 97, 10,096-10,100.

Carbone, G. J., & Bramante, P. D. (1995). Translating monthly temperature from regional to local scale in the southeastern united states, *Climate Research*, 5, 229-242.

Cohen, A. C. J. (1959). Simplified estimators for the normal distribution when samples are singly censored or truncated. *Technometrics*, 1, 217-237.

Conway, D., Wilby, R. L., & Jones, P. D. (1996). Precipitation and air flow indices over the british isles. *Climate Research*, 7, 169-183.

Cowpertwait, P. S. P. (1991). Further developments of the Neyman-Scott clustered point process for modeling rainfall. *Water Resources Research*, 27, 1431-1438.

Cowpertwait, P. S. P. (1995). A generalized spatial-temporal model of rainfall based on a clustered point process. *Proceedings: Mathematical and Physical Sciences*, 450, 163-175.

Cowpertwait, P. S. P., Kilsby, C. G., & Connell, P. E. O. (2002). A space-time Neyman-Scott model of rainfall: Empirical analysis of extremes. *Water Resources Research*, 38, 2001WR000,709.

Cox, D. R., & Isham, V. (1988). A simple spatial-temporal model of rainfall. *Proceedings of the Royal Society London*, p. A415.

Crane, R. G., & Hewitson, B. C. (1998). Doubled CO2 precipitation changes for the Susquehanna basin: down-scaling from the genesis general circulation model. *International Journal of Climatology*, 18, 65-76.

Duckstein, L., Bardossy, A. & Bogardi, I. (1993). Linkage between the occurrence of daily atmospheric circulation patterns and floods: an arizona case study. *Journal of Hydrology*, 143, 413-428.

Favre, A. C., Adlouni, S. E., Perreault, L., Thiemonge, N., and Bobee, B. (2004). Multivariate hydrological frequency analysis using copulas. *Water Resources Research,* 40,2003WR002,456.

Feller, W. (1970). An introduction to probability theory and its applications, Wiley, New York.

Foufoula-Georgiou, E., & Lettenmaier, D. P. (1987). A Markov renewal model for rainfall occurrences. *Water Resources Research*, 23, 875-884.

Fowler, H. J., Kilsby, C. G., & Connell, P. E. (2000). A stochastic rainfall model for the assessment of regional water resource systems under changed climatic conditions. *Hydrological and Earth systems Science*, 4, 263-282.

Gabriel, K. R., & Neumann, J. (1962). A Markov chain model for daily rainfall occurrence at Tel Aviv, 88.

Geng, S., de Vries, F. P., & Supit, I. (1986). A simple method for generating daily rainfall data, *Agric For Meteorol* , 36 , 363-376.

Giorgi, F. (1990). Simulation of regional climate using a limited area model nested in a general circulation model. *Journal of Climate*, 3, 941-963.

Giorgi, F., & Bates, G. T. (1989). The climatological skill of a regional model over complex terrain. *Monthly Weather Review*, 117, 2325-2347.

Giorgi, F., & Mearns, L. O. (1991). Approaches the simulation of regional climate change: review. *Reviews of Geophysics*, 29, 191-216.

Giorgi, F., Hewitson, B., Hulme, J. H. C. M., et al. (2001). Regional climate information-evaluation and projections, in Climate Change 2001 The Scientific Basis, In: Houghton, J.T., Ding, Y., Griggs, D.J., Noguer, M, van der Linden, P.J., Dai, X., Maskell, K., Johnson, C.A. (Eds), Climate change 2001: the scientific basis. Contribution of Working Group 1 to the Third Assessment Report of the Intergov- ernmental Panel on Climate Change, Cambridge University Press, Cambridge UK.

Gregory, J. M., Wigley, T. M. L., & Jones, P. D. (1993). Application of Markov models to area-average daily precipitation series and interannual variability in seasonal totals. *Climate Dynamics*, 8, 299-310.

Grotch, S. L., & MacCracken, M. C. (1991). The use of general circulation models to predict regional climate change. *Journal of Climate*, 4, 286-303.

Harrold, T. I., Sharma, A., & Sheather, S. (2005). A nonparametric model for stochastic generation of daily rainfall amount. *Water Resources Research*, 39, WR002,570.

Hay, L. E., McCabe, G. J., Wolock, D. M., & Ayers, M. A. (1991). Simulation of precipitation by weather type analysis. *Water Resources Research*, 27, 493-501.

Hayhoe, H. N., & Stewart, D. W. (1996). Evaluation of CLIGEN and WXGEN weather data generators under Canadian conditions. *Canadian Water Resources Journal*, 21, 53-57.

Hewitson, B. C., & Crane, R. G. (1992). Large-scale atmospheric controls on local precipitation in tropical Mexico. *Geophysical Research Letters*, 19, 1835-1838.

Hewitson, B. C., & Crane, R. G. (1996). Climate downscaling: techniques and application. *Climate Research*, 7, 85-95.

Hostetler, S. W., & Giorgi, F. (1992). Use of a regional atmospheric model to simulate lake atmosphere feedbacks associated with Pleistocene lakes Lahontan and Bon-Neville. *Climate Dynamics*, 7, 39-44.

Hughes, J., Lettenmaier, D. & Guttorp, P. (1993). A stochastic approach for assessing the effects of changes in regional circulation patterns on local precipitation. *Water Resources Research*, 29, 3303-3315.

Hughes, J. P., & Guttorp, P. (1994a). A class of stochastic models for relating synoptic atmospheric patterns to regional hydrologic phenomena. *Water Resources Research*, 30, 1535-1546.

Hughes, J. P., & Guttorp, P. (1994). Incorporating spatial dependence and atmospheric data in a model of precipitation. *Journal of Applied Meteorology*, 33, 1503-1515.

Hughes, J. P., Guttorp, P., & Charles, S. P. (1999). A non-homogeneous hidden Markov model for precipitation occurrence. *Applied Statistics*, 48, 15-30.

Hulme, M., Briffa, K. R., Jones, P. D., & Senior, C. A. (1993). Validation of GCM control simulations using indices of daily airflow types over the British isles. Climate Dynamics, 9, 95-105.

Hutchinson, M. (1995). Stochastic space-time weather models from ground-based data. *Agricultural and Forest Meteorology*, 73, 237-264.

Hutchinson, M. F., Richardson, C. W., & Dyke, P. T. (1993). Normalization of rainfall across different time steps.

IPCC (2007). Climate Change 2007: The Scientific Basis, In J.T. Houghton and Y. Ding and D. J. Griggs and M. Noguer and P. J. Van Der Linden and X. Dai and K. Maskell and C.A. Johnson (Eds), Contribution of Working Group 1 to the Fourth Assessment report of the Integovernmental Panel on Climate Change, Cambridge University Press, Cambridge, UK.

Johannesson, T., Jonsson, T., Klen, E., & Kaas, E. (1995). Climate change scenarios for nordic countries. *Climate Research*, 5, 181-195.

Jones, P. D., Hulme, M., & Briffa, K. R. (1993). A comparison of lamb circulation types with an objective classification scheme. *International Journal of Climatology*, 13, 655-663.

Jones, R. G., Murphy, J. M., & Noguer, M. (1995). Simulation of climate change over Europe using a nested regional climate model. i. Assessment of control climate, including sensitivity to location of lateral boundaries. *Quarterly Journal of the Royal Meteorological Society*, 121, 1413-1450.

Jothityangkoon, C., Sivapalan, M., & Viney, N. R. (2000). Tests of a space-time model of daily rainfall in Southwestern Australia based on nonhomogeneous random cascades. *Water Resources Research*, 36, 267-284.

Kaas, E. (1993). Greenhouse induced climate change in the Nordic countries as simulated with the Hamburg climate model, Part 2. Statistical interpretation. Scientific Report 93-3, Danish Meteorological Institute, Copenhagen.

Kaas, E. (1994). Greenhouse induced climate change in the Nordic countries as simulated with the Hamburg climate model, An update of statistically interpreted greenhouse induced climate change in the Nordic countries (informal report), Danish Meteorological Institute, Copenhagen.

Katz, R. W. (1977). Precipitation as a chain-dependent process. *Journal of Applied Meteorology*, 16, 671-676.

Katz, R. W. (1996). Use of conditional stochastic models to generate climate change scenarios. *Climatic Change*, 32, 237-255.

Kim, J. W., Chang, J. T., Baker, N. L., Wilks, D. S., & Gates, W. L. (1984). The statistical problem of climate inversion: determination of the relationship between local and large-scale climate. *Monthly Weather Review*, 112, 2069-2077.

Lall, U., Rajagopalan, B., & Tarboton, D. G. (1996). A nonparametric wet/dry spell model for resampling daily precipitation. *Water Resources Research*, 32, 2803-2823.

Lamb, H. H. (1972). British isles weather types and a register of daily sequence of circulation patterns. *Geophysical Memoir*, 116, 1861-1971.

Lambert, M. F., Whiting, J. P., & Metcalfe, A. V. (2003). A non-parametric hidden Markov model for climate state identification. *Hydrology and Earth System Sciences*, 7, 652-667.

McCabe, G. J., & Legates, D. R. (1992). General circulation model simulations of winter and summer sea-level pressures over North America. *International Journal of Climatology*, 32, 815-827.

Mearns, L. O., Rosenzweig, C., & Goldberg, R. (1996). The effect of changes in daily and interannual climatic variability on ceres-wheat: a sensitivity study. *Climatic Change*, 32, 257-292.

Mehrotra, R., & Sharma, A. (2007). A semi-parametric model for stochastic generation of multi-site daily rainfall exhibiting low-frequency variability. *Journal of Hydrology*, 335, 180-193.

Moses, T., Kiladis, G. N., Diaz, H. F., & Barry, R. G. (1987). Characteristics and frequency of reversals in the mean sea level pressure in the North Atlantic sector and their relationship to long-term temperature trends. *Journal of Climatology*, 7, 1330.

Murphy, J. M. (1999). An evaluation of statistical and dynamical techniques for downscaling local climate. *Journal of Climate*, 12, 2256-2284.

Nelsen, R. B. (1999). An Introduction to Copulas, Springer, New York.

Ng, W. W., Panu, U. S., & Lennox, W. C. (2007). Chaos based analytical techniques for daily extreme hydrological observations. *Journal of Hydrology*, 342, 17-41.

Ng, W. W., Panu, U. S., & Lennox, W. C. (2009). Comparative studies in problems of missing extreme daily streamflow records. *Journal of Hydrological Engineering*, 14, 91-100.

Ng, W. W. & Panu, U.S. (2010). Comparisons of traditional and novel stochastic models for the generation of daily precipitation occurrences. *Journal of Hydrology*, 380, 222-236.

Pachter, L., Batzoglou, S., Spitkovsky, V. I., Banks, E., Lander, E. S., Kleit-man, D. J., & Berger, B. (1999). A dictionary-based approach for gene annotation. *Journal of Computational Biology*, 6, 419-430.

Panu, U. S., Ng, W. W., & Lennox, w. C. (2003). Identification and detection of outliers in annual flood flow series, Proceedings of the 16th Canadian Hydrotechnical Conference (CSCE), HY-2:, Burlington,Canada.

Perica, S., & Foufoula-Georgiou, E. (1996). Linkage of scaling and thermodynamic parameters of rainfall: results from midlatitude mesoscale convective systems. *Journal of Geophysical Research*, 101, 7431-7448.

Pickering, N. B., Hansen, J. W., Jones, J. W., Wells, C. M., Chan, V. K., & Godwin, D. C. (1994). Weatherman: a utility for managing and generating daily weather data. Agron J, 86, 332-337.

Racsko, P., Szeidl, L., & Semenov, M. (1991). A serial approach to local stochastic weather models. *Ecological Modelling*, 57, 27-41.

Rajagopalan, B., & Lall, U. (1999). A *k*-nearest-neighbor simulator for daily precipitation and other weather variables. *Water Resources Research*, 35, 3089-3101.

Rasmussen, P. F., & Akintug, B. (2004). Drought frequency analysis with a hidden state Markov model. Proceedings of the 2004 World Water and Environmental Resources 129 Congress: Critical Transitions in Water and Environmental Resources Management, pp. 2078-2087.

Richardson, C. W. (1981). Stochastic simulation of daily precipitation, temperature, and solar radiation. *Water Resources Research*, 17, 182-190.

Richardson, C. W., & Wright, D. A. (1984). WGEN: a model for generating daily weather variables. U.S. Department of Agriculture, *Agricultural Research Service*, ARS-8, p.83.

Rolda'n, J., & Woolhiser, D. A. (1982). Stochastic daily precipitation models, 1, a comparison of occurrence processes. *Water Resources Research*, 18, 1451-1459.

Rummukainen, M. (1997). Methods for statistical downscaling of GCM simulation. SWE-CLIM report, Rossby Centre, SMHI, Norrking.

Salas, J. D., & Boes, D. C. (1980). Shifting level modelling of hydrologic time series. *Advances in Water Resources*, 3, 59-63.

Salas, J. (1993). Analysis and modeling of hydrologic time series, in Handbook of Hydrology, edited by D. Maidment, McGraw-Hill, New York.

Sanso, B., & Guenni, L. (2004). A Bayesian approach to compare observed rainfall data to deterministic simulations. EnvironMetrics, 15, 597-612.

Sansom, J., & Thomson, P. J. (1998). Detecting components in censored and truncated meteorological data. *Environmetrics*, 9, 673-688.

Santer, B., & Wigley, T. M. L. (1990). Regional validation of means, variances, and spatial patterns in general circulation model control runs. Journal of Geophysical Research, 95, 829-850.

Scheuer, E. M., & Stoller, D. S. (1962). On the generation of Normal random vectors. *Ecological Modelling*, 4, 278-281.

Schulze, E. E. (1997). Impacts of global climate change in a hydrologically vulnerable region: challenges to South African hydrologists. *Progress in Physical Geography*, 21, 113-136.

Semenov, M. A., & Barrow, E. M. (1997). Use of a stochastic weather generator in the development of climate change scenarios. *Climatic Change*, 35, 397-414.

Semenov, M. A., & Brooks, R. J. (1999). Spatial interpolation of the LARS-WG stochastic weather generator in Great Britain. *Climate Research*, 11, 137-148.

Semenov, M. A., Brooks, R. J., Barrow, E. M., & Richardson, C. W. (1998). Comparison of the WGEN and LARS-WG stochastic weather generators for diverse climate. *Climate research*, 10, 95-107.

Sharma, A., & Lall, U. (1999). A nonparametric approach for daily rainfall simulation. *Mathematics and Computers in Simulation*, 48, 361-371.

Sklar, A. (1959). Fonctions de re'partition a n-dimensions et leurs marges, vol. 8, Publications de l'Institut de Statistique de L'Universite' de Paris, Paris France.

Smith, E. R., & Schreiber, H. A. (1973). Point processes of seasonal thunderstorm rainfall, 1, distribution of rainfall events. *Water Resources Research*, 9, 871-884.

Smith, J. A. (1987). Statistical modelling of daily rainfall occurrences. *Water Resources Research*, 23, 885-893.

Smith, J. A., & Karr, A. F. (1985). Parameter estimation for a model of space-time rainfall. *Water Resources Research*, 21, 1251-1257.

Smith, R. E., & Schreiber, H. A. (1974). Point processes of seasonal thunderstorm rainfall, 2, rainfall depth probabilities. *Water Resources Research*, 10, 418-423.

Sowden, I. P., & Parker, D. E. (1981). A study of climatic variability of daily central England temperatures in relation to the lamb synoptic types. *Journal of Climatology*, 1, 310.

Srikanthan, R., & McMahon, T. (2001). Stochastic generation of annual, monthly and daily climate data: A review. *Stochastic Hydrology and Earth System Sciences*, 5, 653-670.

Stehlik, J., & Bardossy, A. (2002). Multivariate stochastic downscaling model for generating daily precipitation series based on atmospheric circulation. *Journal of Hydrology*, 256, 120-141.

Stern, R. D., & Coe, R. (1984). A model fitting analysis of daily rainfall data. *Journal of Royal Statistical Society*, Series A, 147, 1-34.

Thyer, M., & Kuczera, G. (2000). Modelling long-term persistence in hydro-climatic time series using a hidden state Markov model. *Water Resources Research*, 36 , 3301-3310.

Thyer, M., & Kuczera, G. (2003a). A hidden Markov model for modelling long-term persistence in multi-site rainfall time series 1. Model calibration using a Bayesian approach. *Journal of Hydrology*, 275, 12-26.

Thyer, M., & Kuczera G. (2003b). A hidden Markov model for modelling long-term persistence in multi-site rainfall time series 2. Real data analysis. *Journal of Hydrology*, 275, 27-48.

von Storch, H. (1999). On the use of inflation in statistical downscaling. *Journal of Climate*, 12, 3505-3506.

von Storch, H., Zorita, E., & Cubasch, U. (1993). Downscaling of global climate change estimates to regional scales: an application to Iberian rainfall in wintertime. *Journal of Climate*, 6, 1161-1171.

Wigley, T. M. L., Jones, P. D., Briffa, K. R., & Smith, G. (1990). Obtaining sub-grid scale information from coarse-resolution general circulation model output. *Journal of Geophysical Research*, 95, 1943-1953.

Wilby, R. L. (1993). The influence of variable weather patterns on river water quantity and quality regimes. *International Journal of Climatology*, 13, 447-459.

Wilby, R. L. (1994). Modelling the relative impact of weather, landuse and groundwater abstraction on lowflows, Research and Development Note No. 268, National Rivers Authority, Bristol UK.

Wilby, R. L. (1995). Simulation of precipitation by weather pattern and frontal analysis. *Journal of Hydrology*, 173, 91-109.

Wilby, R. L. (1997). Non-stationarity in daily precipitation series: implications for GCM downscaling using atmospheric circulation indices. *International Journal of Climatology*, 17, 439-454.

Wilby, R. L., & Wigley, T. M. L. (1997b). Downscaling general circulation model output: a review of methods and limitation. *Progress in Physical Geography*, 21, 530-548.

Wilby, R. L., Charles, S. Zorita, E., Timbal, B., Whetton, P., and Mearns, L. (2004). Guidelines for use of climate scenarios developed from statistical downscaling methods, 2004, "supporting material" of the Intergovernmental Panel on Climate Change for consideration by the IPCC at the request of its task group on data and scenario support for impacts and climate analysis (TGICA).

Wilks, D. S. (1989). Conditioning stochastic daily precipitation models on total monthly precipitation. *Monthly Weather Review*, 25, 1429-1439.

Wilks, D. S. (1992). Adapting stochastic weather generation algorithms for climate change studies. *Climatic Change*, 22, 67-84.

Wilks, D. S. (1998). Multisite generalization of a daily stochastic precipitation generation model. *Journal of Hydrology*, 210, 178-191.

Wilks, D. S. (1999a). Multisite downscaling of daily precipitation with a stochastic weather generator. *Climate Research*, 11, 125-136.

Wilks, D. S. (1999b) Interannual variability and extreme-value characteristics of several stochastic daily precipitation models. *Journal of Agricultural Meteorology*, 93, 153-169.

Wilks, D. S. (2008). High-resolution spatial interpolation of weather generator parameters using local weighted regressions. *Agricultural and Forest Meteorology*, 148, 111-120.

Wilks, D. S., & Wilby, R. L. (1999). The weather generation game: A review of stochastic weather models. *Progress in Physical Geography*, 23, 329-357.

Wilson, L. L., & Skyllingstad, D. P. L. E. (1992). A hierarchical stochastic model of large-scale atmospheric circulation patterns and multiple station daily precipitation. *Journal of Geophysical Research*, D3, 2791-2809.

Woolhiser, D. A., & J. Rolda'n. (1982). Stochastic daily precipitation models, 2, a comparison of distributions of amounts. *Water Resources Research*, 18, 1461-1468.

Xu, C. (1999). From GCMs to river flow: a review of downscaling methods and hydrologic modelling approaches. *Progress in Physical Geography*, 23, 229-249.

Yang, D., Goodison, B. E., Ishida, S., & Benson, C. S. (1998). Adjustment of daily precipitation data at 10 climate stations in Alaska: Application of world meteorological organization intercomparison results. *Water Resources Research*, 34, 241-256.

Yoo, C., & Ha, E. (2002). Basin average rainfall and its sampling error. *Water resources research*, 38, 1259-1265.

Young, K. C. (1994). A multivariate chain model for simulating climatic parameters from daily data. *Journal of Applied Meteorology*, 33 , 661-671.

Zorita, E., Kharin, V., & von Storch, H. (1992). The atmospheric circulation and sea surface temperature in the North Atlantic area in winter: their interaction and relevance for Iberian precipitation. *Journal of Climate*, 5, 1097-1108.

Zucchini, W., & Guttorp, P. (1991). A hidden Markov model for space-time precipitation. *Water Resources Research*, 27, 1917-1923.

In: Precipitation: Prediction, Formation, Environmental Impact ISBN: 978-1-62100-447-9
Editors: Henry Dohring and Jeremy Dixon

Chapter 4

PREPARATION METHODS AND INVESTIGATION OF PROPERTIES OF MG–AL LAYERED DOUBLE HYDROXIDES

Tomohito Kameda*[*] *and Toshiaki Yoshioka
Graduate School of Environmental Studies, Tohoku University, Aramaki, Aoba-ku, Sendai, Japan

ABSTRACT

Magnesium–aluminum layered double hydroxides (Mg–Al LDHs) possess anion-exchange properties that can potentially be utilized for water preservation and purification. Therefore, a simple and convenient method for preparing Mg–Al LDHs is required. In this chapter, we developed new preparation methods of Mg–Al LDH by using seawater and $Ca(OH)_2$ as magnesium and alkali resources. Furthermore, dolomite and polyaluminum chloride were examined as potential magnesium, aluminum, and alkali resources for the production of Mg–Al LDHs. In addition, we analyzed the formation of Mg–Al LDH, as below. Mg–Al solution, together with alkali (used for pH control), was added dropwise to deionized water. In the local environment, where the droplets of alkali were added, the pH was very high. This high pH resulted in the formation of $Al(OH)_4^-$ from $Al(OH)_3$; this $Al(OH)_3$ formed in solution with the addition of Mg–Al solution. The $Al(OH)_4^-$ quickly reacted with Mg^{2+} and other anions present in the solution, thereby leading to the formation of Mg–Al LDH. Moreover, we examined the characteristics of anion intercalation during Mg–Al LDH precipitation and investigated the Mg–Al LDH particle properties for the various preparation methods.

INTRODUCTION

Magnesium–aluminum layered double hydroxide (Mg–Al LDH) is an inorganic compound possessing anion exchange properties [1, 2]. Mg–Al LDH is represented by the

[*] E-mail address: kameda@env.che.tohoku.ac.jp.

general formula $[Mg^{2+}{}_{1-x}Al^{3+}{}_{x}(OH)_2](A^{n-})_{x/n}{\bullet}mH_2O$, where x denotes an Al/(Mg + Al) molar ratio ($0.20 \leqq x \leqq 0.33$) and A^{n-} is an anion with a valency of n [3, 4]. Mg–Al LDH consists of a stack of Al-bearing brucite-like octahedral layers. The positive charge accrued, due to the replacement of Mg with Al, is electrically neutralized by the interlayer anions. The interlayer space unoccupied by the intercalated anions is occupied by water molecules. CO_3^{2-} anion-intercalated Mg–Al LDH (CO_3•Mg–Al LDH) has been used as an antacid and a flame retardant [2]. Mg–Al LDH has excellent heat and radiation resisting properties, surpassing its properties as an ion exchange resin; because of its ion exchange properties, it can potentially remove toxic anionic species from wastewater. In order to confirm this ability to remove toxic anions, we treated mineral acids, chlorides, fluorides, borates, and tetrafluoroborates using calcined Mg–Al LDH [5−12]. Furthermore, Mg–Al LDHs, modified with organic acid anions, can absorb heavy metal ions and nonionic organic compounds from aqueous solutions [13−21]. In recent times, Mg–Al LDH has been produced using general reagents at temperatures above 80°C, and further purified [22]. However, this process is unsuitable for use in wastewater treatment because a large amount of Mg–Al LDH (not obtained by this method) is required for the preservation of water environments. In order to preserve and purify water, a simple and convenient method for preparing Mg–Al LDHs is required.

In the laboratory, Mg–Al LDH is usually prepared by mixing alkali solutions, such as NaOH with Mg–Al solutions, such as $MgCl_2$−$AlCl_3$ and $Mg(NO_3)_2$−$Al(NO_3)_3$ [23−25]. A few studies have reported methods of producing commercial-scale Mg–Al LDH. Some researchers have examined $Mg(OH)_2$, $MgCO_3$, MgO, $NaAlO_2$, and $Al(OH)_3$ as potential magnesium and aluminum resources for the production of Mg–Al LDHs [26−28]. However, these preparation methods require lengthy reaction times (7 to 15 hours) at temperatures of 95–100°C in order to obtain pure Mg–Al LDH crystals without any by-products. Therefore, we have developed a method of preparing Mg–Al LDH using new sources, under mild production conditions. In this chapter, seawater, dolomite ($CaMg(CO_3)_2$), polyaluminum chloride (PAC), $Ca(OH)_2$, and Na_2CO_3 were examined as potential magnesium, aluminum, and alkali resources for the preparation of Mg–Al LDH. Furthermore, we studied the characteristics of anion intercalation during Mg–Al LDH precipitation, and we investigated the varying particle properties of Mg–Al LDH, as depending on the preparation methods.

PREPARATION FROM SEAWATER

Soda Ash Process

Artificial seawater and Na_2CO_3 were examined as potential magnesium and alkali resources [29]. The stoichiometric equation, governing the preparation of Mg–Al LDH by the soda ash process, can be described by equation (1) as follows:

$$(1-x)Mg^{2+} + xAl^{3+} + 2OH^{-} + (x/2)Na_2CO_3 \rightarrow Mg_{1-x}Al_x(OH)_2\,(CO_3)_{x/2} + xNa^{+} \quad (1)$$
$$(0.20 \leqq x \leqq 0.33)$$

Our artificial seawater sample contained 0.41 M NaCl, 0.028 M Na_2SO_4, 0.053 M $MgCl_2$, and 0.010 M $CaCl_2$. In this case, Ca^{2+} ions must be removed from artificial seawater as a

$CaCO_3$ precipitate, an impurity that co-precipitates with Mg–Al LDH when a Na_2CO_3 solution is added to artificial seawater containing $AlCl_3$. In our method, Ca^{2+}-free artificial seawater was obtained by adding a mixed 0.1 M Na_2CO_3–0.2 M $NaHCO_3$ solution to the artificial seawater until the 1.2 equivalent of Ca^{2+} in the artificial seawater, and stirring at 95°C for 1 hour [30]. In this case, 3% and 96% of Mg^{2+} and Ca^{2+} were precipitated, respectively, suggesting that Ca^{2+} was selectively removed. A 1.0 M Na_2CO_3 solution was added to 500 ml of the Ca^{2+}-free artificial seawater containing $AlCl_3$ (with initial Mg/Al molar ratios of 2.0–3.7) until a pH of 10 was obtained; subsequently, the solution was stirred at 60°C for 1 hour. The precipitates were filtered, washed with deionized water, and dried at 105°C for 24 hour.

From the X-ray diffraction (XRD) patterns of the precipitates, the Mg–Al LDHs were found to have precipitated without by-products. In the initial Mg/Al molar ratios of 2.0–2.3, the percentage value of each recovery of Mg^{2+} and Al^{3+} from the solution was above 98%. The basal spacing value, d(003), of the precipitate obtained from the initial Mg/Al molar ratio of 2.1 was 7.72 Å; this value was identical to the spacing value of the CO_3•Mg–Al LDH (7.70 Å) reported by Sato et al. [31]. This suggests that CO_3^{2-} was intercalated in the interlayer of the precipitated Mg–Al LDH.

Milk of Lime Process

Artificial seawater and $Ca(OH)_2$ were examined as magnesium and alkali resources [29, 32]. The stoichiometric equation governing the preparation of Mg–Al LDH by the milk of lime process can be described by equation (2) as follows:

$$(1-x)Mg^{2+} + xAl^{3+} + Ca(OH)_2 + (x/n)A^{n-} \rightarrow Mg_{1-x}Al_x(OH)_2\,(A^n)_{x/n} + Ca^{2+} \qquad (2)$$
$$(0.20 \leqq x \leqq 0.33)$$

A $Ca(OH)_2$ slurry of 0.7 wt% was added to 500 ml of artificial seawater containing $AlCl_3$ (with an initial Mg/Al molar ratio of 1.0–5.0) until a pH of 10.5 was obtained, and stirred at 60°C for 1 hour. The precipitates were filtered, washed with deionized water, and dried at 105°C for 24 hours.

Figure 1 shows the XRD patterns of the precipitates obtained from artificial seawater containing $AlCl_3$. The XRD peaks of the precipitates were attributed to the presence of hydrotalcite (JCPDS card 22-700), a naturally occurring hydroxycarbonate of magnesium and aluminum ($Mg_6Al_2(OH)_{16}CO_3{\cdot}4H_2O$), thereby indicating the precipitation of Mg–Al LDH. In particular, Mg–Al LDH was precipitated without by-products, with initial Mg/Al molar ratios of 2.0–4.0. Figure 2 shows the effect of the initial Mg/Al molar ratio in the artificial seawater containing $AlCl_3$ on the Mg/Al molar ratio in the precipitates, and the precipitated Mg^{2+} and Al^{3+}. When the initial Mg/Al molar ratio was 2.4–3.0, each recovery of Mg^{2+} and Al^{3+} from the solution was quantitatively precipitated. A mixture of Mg–Al LDH and $Al(OH)_3$ was precipitated for an initial Mg/Al molar ratio of 1.0, and a mixture of Mg–Al LDH and $Mg(OH)_2$ was precipitated for initial Mg/Al molar ratios in the range of 4.5–5.0. Even at 25°C, Mg–Al LDH was quantitatively precipitated from the solution with an initial Mg/Al molar ratio of 3.0. Further, Mg–Al LDH was precipitated from the solution at 25°C, for an initial Mg/Al molar ratio of 3.0 when 0.7–10.0 wt% $Ca(OH)_2$ slurry and $Ca(OH)_2$ powder were used. Mg^{2+} and Al^{3+} were quantitatively precipitated from the solution.

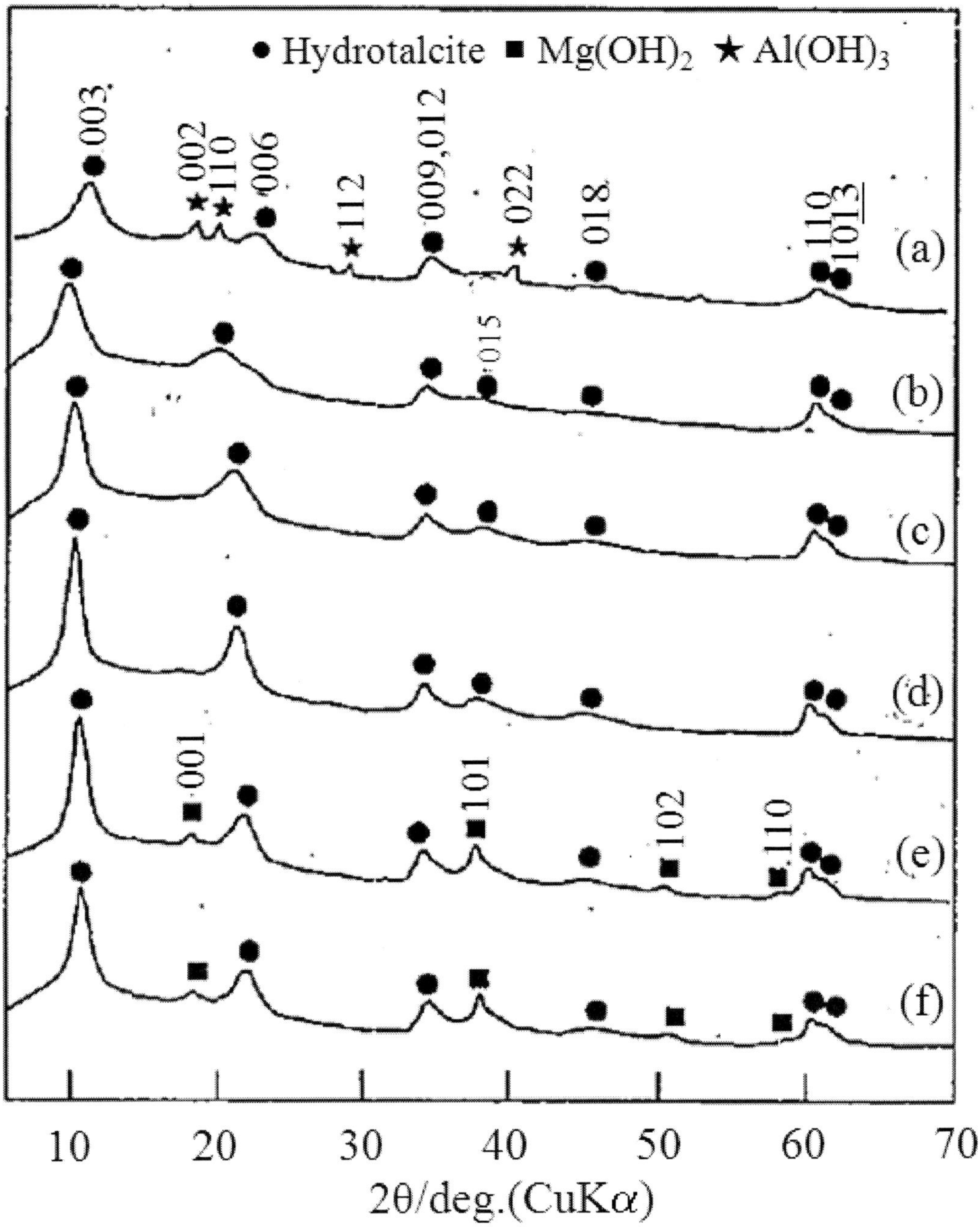

Initial Mg/Al molar ratios were (a) 1.0, (b) 2.0, (c) 3.0, (d) 4.0, (e) 4.5, and (f) 5.0.

Figure 1. X-ray diffraction (XRD) patterns of precipitates from artificial seawater containing $AlCl_3$.

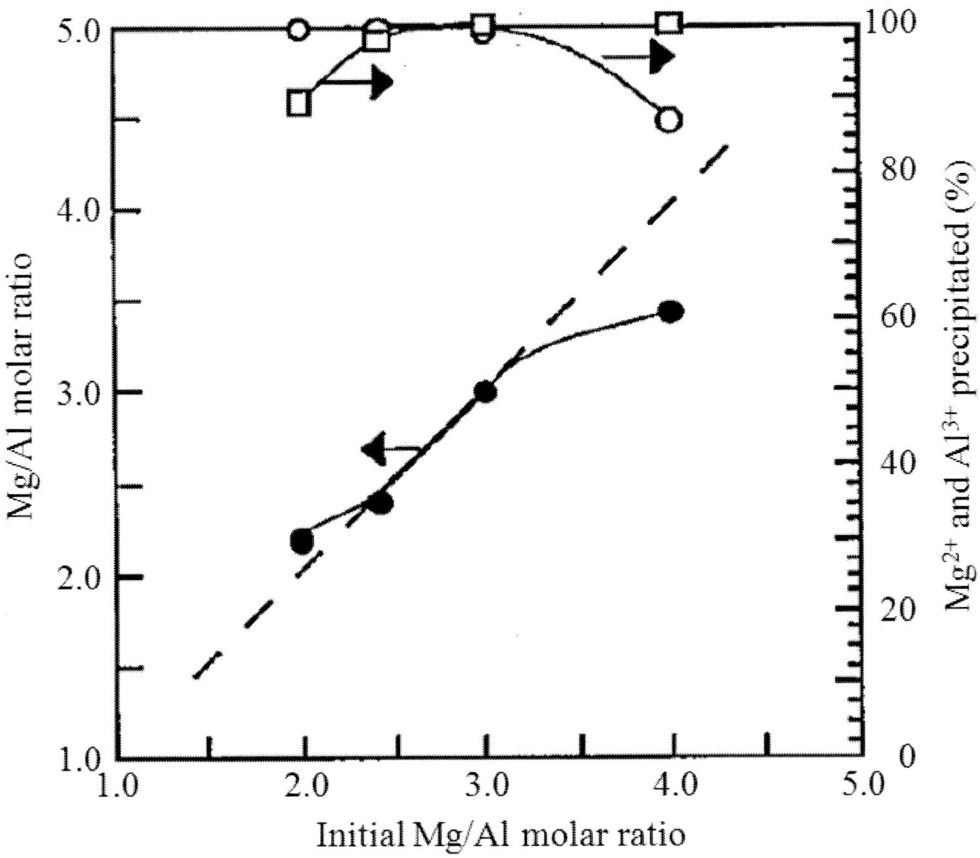

Figure 2. Effect of initial Mg/Al molar ratio in artificial seawater containing $AlCl_3$ on Mg/Al molar ratio in precipitates (●), and Mg^{2+} (○) and Al^{3+} (□) precipitated.

Furthermore, natural seawater was examined as a magnesium resource. Natural seawater contained 424 mM Na^+, 9 mM K^+, 48 mM Mg^{2+}, 11 mM Ca^{2+}, 25 mM SO_4^{2-}, 500 mM Cl^-, and 1 mM Br^-. A $Ca(OH)_2$ slurry of 10.0 wt% was added to 500 ml of natural seawater containing $AlCl_3$ (with an initial Mg/Al molar ratio of 3.0) until a pH of 10.5 was obtained, and stirred at 25°C for 1 hour.

Mg–Al LDH was also precipitated as a single phase from natural seawater, as well as from artificial seawater. Mg^{2+} and Al^{3+} were quantitatively precipitated in both cases. The precipitated Mg–Al LDH contained 8.5 wt% SO_4^{2-} and 1.2 wt% Cl^-. The basal spacing value, d(003) of the Mg–Al LDH, was 8.20 Å; this value was close to the spacing value of the SO_4•Mg–Al LDH (8.60 Å) reported by Sato et al. [31]. This suggests that SO_4^{2-} was intercalated in the prepared Mg–Al LDH. The chemical composition of the precipitate was calculated to be $[Mg_{0.75}Al_{0.25}(OH)_2][(SO_4)_{0.06}(Cl)_{0.02}(OH)_{0.11}{}^*]\bullet 0.27H_2O^*$ (*Balance). In order to compensate for the electric charge of the LDH host layer, the OH^- in the solution was intercalated in the interlayer of the Mg–Al LDH as another anion.

PREPARATION FROM DOLOMITE

Use of $AlCl_3$

We examined the possibility of preparing Mg–Al LDH by using calcined dolomite as a potential magnesium and alkali source [33]. Calcined dolomite comprises a mixture of CaO and MgO. Theoretically, the preparation of Mg–Al LDH, using calcined dolomite, is governed by equation (3), as given below. Further, we expected the use of calcined dolomite to further reduce the amount of Mg^{2+} in solution by 62.5–75%.

$$(0.5-x)Mg^{2+} + xAl^{3+} + 0.5MgO + 0.5CaO + H_2O + (x/n)A^{n-}$$
$$\rightarrow Mg_{1-x}Al_x(OH)_2\,(A^n)_{x/n} + 0.5Ca^{2+} \quad (3)$$
$$(0.20 \leqq x \leqq 0.33)$$

The calcination of dolomite ($CaMg(CO_3)_2$) was carried out in an electric furnace at 1,000°C for 1 hour. The CaO and MgO content in the calcined dolomite were determined to be 59.5% and 39.8%, respectively. $MgCl_2$ was added to 100 ml of 0.1 M $AlCl_3$ solution in $(Mg)_{Cl}$/Al molar ratios of (i) 0.5, (ii) 1.0, and (iii) 1.5 at 25°C and stirred. The calcined dolomite was added to the mixed solution in $(Mg)_D$/Al molar ratios of (i) 1.5, (ii) 2.0, and (iii) 2.5 in a burst process; $(Mg)_{Cl}$: Mg was obtained from $MgCl_2$ and $(Mg)_D$: Mg was obtained from the calcined dolomite. The slurry prepared with $((Mg)_D + (Mg)_{Cl})$/Al molar ratios of (i) 2.0, (ii) 3.0, and (iii) 4.0 was stirred at 25°C and 90°C for a duration of 1–16 hours. The obtained precipitates were filtered, washed with water, and dried at 105°C for 24 hours.

After stirring the mixture at 25°C for 1 hour, Mg–Al LDH was precipitated as a single phase at $((Mg)_D + (Mg)_{Cl})$/Al molar ratios of 2.0 and 3.0; in addition, a small amount of $Mg(OH)_2$ was precipitated as a by-product at the $((Mg)_D + (Mg)_{Cl})$/Al molar ratio of 4.0. It has previously been determined that Mg–Al LDH is formed by the reaction of the dissolved Mg and Al species [34]. For the reactions occurring at molar ratios of 2.0 and 3.0, the measured pH values were 6.8 and 9.0, respectively. Thus, the formation of Mg–Al LDH was attributed to the reaction of the Mg species dissolved from the calcined dolomite, and the Al species from $AlCl_3$. In contrast, it is speculated that at a molar ratio of 4.0, the rate of hydration of MgO was faster than the formation of Mg–Al LDH. This is supported by the fact that the pH value measured after the reaction was 10.4; the pH value increased because the amount of the calcined dolomite added at this ratio was larger than that at molar ratios of 2.0 and 3.0. At $((Mg)_D + (Mg)_{Cl})$/Al molar ratios of 2.0 and 3.0, an almost quantitative precipitation of Mg and Al occurred, while 4 wt% of Cl^- remained in both samples of Mg–Al LDHs obtained at these molar ratios. The basal spacing values, d(003) of the Mg–Al LDHs, prepared at the $((Mg)_D + (Mg)_{Cl})$/Al molar ratios of 2.0 and 3.0 were 7.91 Å and 7.96 Å, respectively; these values were close to the spacing value of the Cl•Mg–Al LDH (7.70 Å) reported by Sato et al. [31]. This suggests that Cl^- was intercalated in both samples of Mg–Al LDHs obtained at the molar rations of 2.0 and 3.0. For the $((Mg)_D + (Mg)_{Cl})$/Al molar ratio of 4.0 at 90°C, the XRD peak intensity ratios of $Mg(OH)_2$ to Mg–Al LDH decreased with time. It is assumed that acid was released by the hydrolysis of amorphous aluminum hydroxide in the precipitate with time, and $Mg(OH)_2$ dissolved gradually. Accordingly, Mg–Al LDH was gradually regenerated by the reactions of the dissolved Mg and Al species. To date, there have been no reports of Mg–Al LDH preparation, as a single phase using salts of low solubility (e.g., $Mg(OH)_2$) at ambient temperature. It is noteworthy that Mg–Al LDH could be produced as a single phase at 25°C by the addition of calcined dolomite to mixtures of $MgCl_2$ and $AlCl_3$ at $((Mg)_D + (Mg)_{Cl})/Al$ molar ratios of 2.0 and 3.0.

Use of Polyaluminum Chloride

Our previous method used $AlCl_3$ as the aluminum source. However, because $AlCl_3$ is a reagent that requires purification, a more convenient source is required. In this chapter, we investigated the use of a polyaluminum chloride (PAC) solution to obtain Mg–Al LDHs; PAC has previously been used in the coagulation sedimentation process during water treatment [35]. Further, PAC is often used in industrial-scale applications. In addition, it is relatively less expensive than other aluminum salts such as $AlCl_3$.

The concentrations of elements contained in the PAC solution were as follows: [Al] = 2.45 M, [Mg] = 0.06 M, [Ca] = 0.13 M, [Cl] = 2.99 M, and [SO_4] = 0.33 M. The composition of the PAC is represented as $AlMg_{0.02}Ca_{0.05}(OH)_{1.64}Cl_{1.22}(SO_4)_{0.14}$. Firstly, we attempted to use a mixed solution of $MgCl_2$ and PAC, with initial Mg/Al molar ratios of 2.0 to 4.0, to obtain the Mg–Al LDHs. This reaction is expressed in Eq. (4) as follows:

$$(z–0.02)Mg^{2+} + (Al^{3+} + 0.02Mg^{2+} + 0.05Ca^{2+}) + 2(z+1)OH^- + 1/nA^{n-}$$
$$\rightarrow Mg_zAl(OH)_{2(z+1)}(A)_{1/n} + 0.05Ca^{2+} \quad z = 2.0, 3.0, 4.0 \qquad (4)$$

In this case, the "$Al^{3+} + 0.02Mg^{2+} + 0.05Ca^{2+}$" component was derived from the PAC. $MgCl_2$ was added to 100 ml of PAC solutions, containing 0.167, 0.125, and 0.1 M of Al^{3+}, in Mg/Al molar ratios of 2.0, 3.0, and 4.0, respectively. The total concentrations of Mg^{2+} and Al^{3+} in each solution were 0.5 M. Subsequently, a 2 M NaOH solution and deionized water were added to the solution while stirring until the pH and the total volume of the solution approached 10.5 and 150 mL, respectively. The mixture was stirred continuously at 25°C for 1 hour; subsequently, it was left at 25°C for 12 hours. The resulting suspension was washed with deionized water and centrifuged repeatedly. The precipitate was dried at 105°C for 24 hours.

Figure 3 shows the XRD patterns of the precipitates prepared from a mixed solution of $MgCl_2$ and PAC with initial Mg/Al molar ratios of 2.0–4.0. Based on the XRD patterns for the precipitates, the peaks at all Mg/Al molar ratios were ascribed to the presence of hydrotalcite (JCPDS card 22-700), a naturally occurring hydroxycarbonate of magnesium and aluminum, formulated as $Mg_6Al_2(OH)_{16}CO_3•4H_2O$. This indicated that the prepared precipitates had the structure of Mg–Al LDH. Table 1 shows the molar ratios for the precipitates of Mg/Al, Cl/Al, and SO_4/Al. The precipitates contained Cl^- and SO_4^{2-} that were derived from $MgCl_2$ and PAC, respectively. Consequently, the precipitates were confirmed as Mg–Al LDHs intercalated with Cl^- and SO_4^{2-}. For all initial Mg/Al molar ratios, the Cl/Al molar ratios were larger than the SO_4/Al molar ratios, i.e., the content of Cl^- was more than that of SO_4^{2-}. The basal spacing values, d(003), of the precipitates were 7.9~8.1 Å; these values were close to the spacing values of the Cl•Mg–Al LDH reported by Sato et al. [31]. This confirmed that the precipitates mainly consisted of Cl•Mg–Al LDH. Consequently, PAC was found to be a potentially useful aluminum source for preparing Mg–Al LDHs. Table 1 shows that the Mg/Al molar ratios of the precipitates did not correspond to the initial ratios of the starting solutions. For initial Mg/Al molar ratios of 2.0 and 3.0, the ratios in the precipitates were larger than the initial values, suggesting that Al^{3+} remained in the solution. In contrast, Mg^{2+} definitely remained in the solution at the initial Mg/Al molar ratio of 4.0 because the ratio in the precipitate was lower than the initial value. These results indicate that the reaction leading to Mg–Al LDH formation probably did not reach equilibrium. Furthermore, the Mg/Al molar ratios in the precipitates tended to be close to 3.0 [29, 32], thereby suggesting that the stable Mg/Al molar ratio in Mg–Al LDH is 3.0. This may have resulted because the Mg/Al molar ratio in natural hydrotalcite is 3.0. From the XRD patterns, it can be noticed that the crystalline nature of the precipitate decreased as the initial Mg/Al molar ratio decreased; this can be attributed to the disordered brucite-like octahedral layer distances of the Mg–Al LDH, accompanied by the increased Al^{3+}content incorporated into this layer. In the case involving $MgCl_2$, PAC, and NaOH, the Mg–Al LDH was most likely formed by the reaction of Mg^{2+} and $Al(OH)_4^-$. For the Al species, $Al(OH)_3$ may be produced due to the dropwise addition of NaOH solution to a mixed solution of $MgCl_2$ and PAC, to a pH of 10.5, because PAC can create polymers of aluminum hydroxides. $Al(OH)_4^-$ is formed by the dissolution of $Al(OH)_3$ due to the high local pH caused by adding the NaOH solution. This $Al(OH)_4^-$ reacted with Mg^{2+}, thereby leading to the Mg–Al LDH formation.

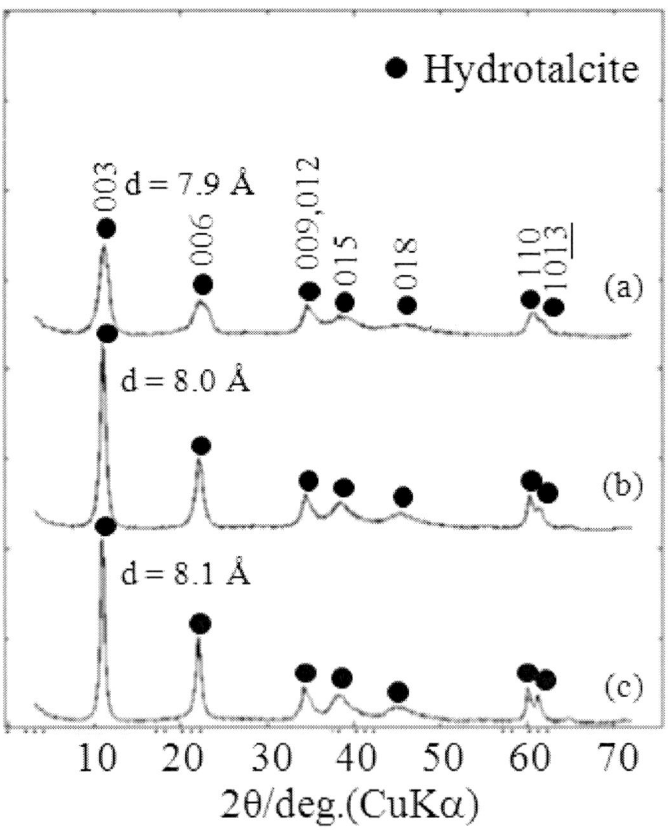

Initial Mg/Al molar ratios were (a) 2.0, (b) 3.0, and (c) 4.0.

Figure 3. XRD patterns of precipitates prepared from a mixed solution of $MgCl_2$ and PAC.

Table 1. Molar ratios of Mg/Al, Cl/Al, and SO_4/Al for precipitates prepared from a mixed solution of $MgCl_2$ and PAC with initial Mg/Al molar ratios of 2.0, 3.0, and 4.0

Initial	Precipitate		
Mg/Al	Mg/Al	Cl/Al	SO_4/Al
2.0	2.71	0.40	0.08
3.0	3.29	0.64	0.08
4.0	3.70	0.66	0.07

Secondly, we attempted to use PAC and calcined dolomite to produce the Mg–Al LDHs; this reaction is expressed in Eq. (5) as follows:

$$(0.5z+0.1175)(CaO{\bullet}MgO) + (0.5z-0.1375)Mg^{2+} + (Al^{3+} + 0.02Mg^{2+} + 0.05Ca^{2+}) + (z+0.235)H_2O + 1.53OH^- + 1/nA^{n-} \rightarrow Mg_zAl(OH)_{2(z+1)}(A)_{1/n} + (0.5z + 0.1675)Ca^{2+} z = 2.0, 3.0, 4.0 \quad (5)$$

In this case, the "CaO•MgO" component was derived from the calcined dolomite. In our procedure, $MgCl_2$ and calcined dolomite were added to 100 ml of PAC solutions containing 0.167, 0.125, and 0.1 M of Al^{3+} in Mg/Al molar ratios of 2.0, 3.0, and 4.0, respectively.

Figure 4 shows the XRD patterns of precipitates prepared from calcined dolomite and a mixed solution of $MgCl_2$ and PAC, with initial Mg/Al molar ratios of 2.0–4.0. For all Mg/Al molar ratios, the XRD patterns indicated the formation of Mg–Al LDH. In particular, $Al(OH)_3$ was also formed for the Mg/Al molar ratios of 3.0 and 4.0. The Mg–Al LDH was produced without by-products only for the initial Mg/Al molar ratio of 2.0, and the molar ratios of Mg/Al, Cl/Al, and SO_4/Al in the formed precipitate were 2.16, 0.24, and 0.13, respectively. The calcined dolomite (CaO•MgO) was immediately hydrated after its addition to the mixed solution of $MgCl_2$ and PAC. Consequently, Mg–Al LDH would be formed by the reaction of $MgOH^+$ and $Al(OH)_4^-$. The $MgOH^+$ is an intermediate product in the hydration process of the MgO component of CaO•MgO. The CaO component is converted into $Ca(OH)_2$ due to the hydration, thereby leading to an increase in the pH of the solution. This results in the production of $Al(OH)_4^-$ through the formation of $Al(OH)_3$ derived from Al^{3+} in the PAC. The production of $Al(OH)_3$ was observed for initial Mg/Al molar ratios of 3.0 and 4.0; these observations may be interpreted as below. With an increasing initial Mg/Al molar ratio, the amount of added CaO•MgO increased, as expressed in Eq. (5); this caused the prompt hydration of MgO. Therefore, the $Al(OH)_4^-$ did not sufficiently react with the $MgOH^+$. Because of the consumption of OH^- in the formation of Mg–Al LDH, the pH of the solution decreases. Accordingly, the remaining $Al(OH)_4^-$ converts into $Al(OH)_3$. The XRD peak intensity ratio of $Al(OH)_3$ to Mg–Al LDH was observed to decrease, with increases in the initial Mg/Al molar ratio, thereby indicating that the precipitate for the initial Mg/Al molar ratio of 4.0 had scarcely contained $Al(OH)_3$. This suggests that the pH of the solution did not decrease substantially until the formation of stable $Al(OH)_3$ [36], due to the large amount of OH^- derived from the CaO•MgO. To summarize, Mg–Al LDH can be prepared with PAC as a new aluminum source at ambient temperature; in addition, Mg–Al LDH can also be prepared by using PAC and calcined dolomite as magnesium and alkali sources.

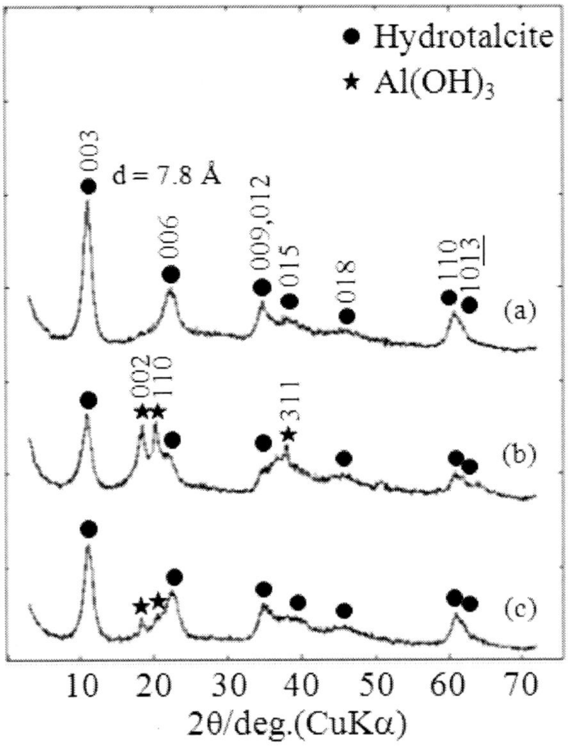

Initial Mg/Al molar ratios were (a) 2.0, (b) 3.0, and (c) 4.0.

Figure 4. XRD patterns of precipitates prepared from calcined dolomite and a mixed solution of $MgCl_2$ and PAC.

Consideration of Mg–Al LDH Precipitation

Formation of Mg–Al LDH

We have speculated that Mg–Al LDH is formed through the reactions of Mg^{2+}, $Al(OH)_4^-$, and anions, as expressed by Eq. (6) [37].

$$(1-x)Mg^{2+} + xAl(OH)_4^- + (2-4x)OH^- + xA^{n-} \rightarrow [Mg_{1-x}Al_x(OH)_2](A^{n-})_{x/n} \quad (6)$$

The Mg-Al solution, together with an alkali (used for pH control), is added dropwise to deionized water. In the local environment where the droplets of alkali are added, the pH is very high. This high pH results in the formation of $Al(OH)_4^-$ from $Al(OH)_3$. This $Al(OH)_3$ forms in solution with the addition of the Mg–Al solution. The $Al(OH)_4^-$ quickly reacts with Mg^{2+} and the anions present, leading to the formation of Mg–Al LDH.

Vanderlaan et al. have studied the formation of Mg–Al LDH during the aging of aluminum hydroxycarbonate gel and magnesium hydroxide gel mixtures [34]. In this case, the aluminum hydroxycarbonate gel was prepared by the addition of an $AlCl_3$ solution to $NaHCO_3$ and Na_2CO_3 solutions. At pH values of 8−9, magnesium hydroxide has a relatively high tendency to donate hydroxyls, as described by the following equation:

$$Mg(OH)_2 \rightarrow MgOH^+ + OH^- \quad (7)$$

Under these conditions, aluminum hydroxide will tend to gain hydroxyls to form aluminates, as described by the following equation:

$$Al(OH)_3 + OH^- \rightarrow Al(OH)_4^- \quad (8)$$

Initially, the pH value of the mixture remains relatively constant due to the balancing action of the reactions shown by Eqs. (7) and (8). After a period of time, the aluminum hydroxide is no longer able to readily assimilate hydroxyls, and the suspension pH begins to rise due to the continued dissolution of magnesium hydroxide. The formation of Mg–Al LDH occurs by the coalescence of the products obtained in Eqs. (7) and (8). This is shown in Eq. (9).

$$Na^+Al(OH)_4^- + 2[(MgOH)^+Cl^-] \rightarrow [AlMg_2(OH)_6]^+Cl^- + NaCl \quad (9)$$

Characteristics of Anion Intercalation

The anion-exchange properties of Mg–Al LDHs are governed by the charge density of the anions, i.e., anions with a higher charge density in solution will replace the anions with a lower charge density in the interlayer of Mg–Al LDHs [1]. Mg–Al LDHs are converted to Mg-Al oxide by calcination at a temperature range of 450–800°C. The Mg-Al oxide is able to rehydrate and combine with the anions in solution to reconstruct the Mg–Al LDH structure. Mg-Al oxides combine more easily with anions that have high charge densities than with those that have low charge densities [7, 31, 38]. Furthermore, Mg-Al oxide compounds, with a low Mg/Al molar ratio, combine with anions having high charge densities [7, 38, 39] because decreasing the Mg/Al molar ratio increases the overall positive charge in the reconstructed Mg–Al LDH. This decreased ratio promotes the intercalation of anions with higher charge densities to compensate for the increased positive charge [40]. It is unclear whether the characteristics of anion intercalation during Mg–Al LDH precipitation are similar to those of a Mg–Al LDH anion-exchange and Mg–Al LDH reconstruction. Lukashin et al. explored the correlation between the solution's pH and the formation of Mg–Al LDHs [41]; however, they did not elucidate the effects of preparation conditions on anion intercalation during Mg–Al LDH formation. In order to comprehensively understand anion intercalation into the interlayer of Mg–Al LDHs, the effects of the preparation conditions must be examined. In this section, we discuss our preparation of Cl•Mg–Al LDH and NO_3•Mg–Al LDH intercalated with Cl^- and NO_3^-, respectively. We examined the effects of the changes in the pH values after the reaction, the concentrations of Cl^- and NO_3^- in the solutions, and the Mg/Al molar ratio of the Mg–Al LDHs in relation to anion intercalation into the interlayer during preparation.

The two Mg–Al solutions had concentrations of $[Mg^{2+} + Al^{3+}] = 0.04–2.5$ M and ratios of Mg/Al = 2–4. These two solutions, i.e., the Mg–Al chloride solution and Mg–Al nitrate solution, were prepared by dissolving a mixture of $MgCl_2•6H_2O$ and $AlCl_3•6H_2O$, and a mixture of $Mg(NO_3)_2•6H_2O$ and $Al(NO_3)_3•9H_2O$, respectively, in deionized water. The Mg-Al solution (100 ml), followed by deionized water (100 ml), was added at a rate of 5 ml/min to 200 ml of deionized water at 25°C, accompanied by vigorous agitation. Simultaneously, 0.5 M NaOH was added to the solution, in a controlled manner to achieve the experimental pH values; these values ranged from 8 to 13. Deionized water was further added to this mixture to obtain a total solution volume of 500 ml. The mixture was stirred continuously for 1 hour at 60°C; subsequently, it was incubated at room temperature for 12 hour. Nitrogen gas was bubbled into the solution throughout the co-precipitation process to minimize the dissolved CO_2. After incubation, the resulting suspension was repeatedly washed with deionized water and collected by centrifugation. The precipitate was dried at 105°C for 24 hours.

Figure 5 shows the effect of the final pH values for the mixture, after the reaction, on the molar ratio of Cl/Al and NO_3/Al in the precipitates; these pH values were obtained in the mixture after 12 hours at room temperature. The terms $[Cl^-]_T$ and $[NO_3^-]_T$ (in Figure 5) are the theoretical concentrations of Cl^- and NO_3^-, respectively, in 500 ml of the prepared solution. Using XRD, the precipitates used for all plots in Figure 5 were confirmed to be Cl•Mg–Al LDH or NO_3•Mg–Al LDH,. The molar ratio of Cl/Al and NO_3/Al in the precipitates relates to the degree of intercalation of these anions into the interlayer of the Mg–Al LDHs. At the $[Cl^-]_T$ value of 0.11 M, the Cl/Al molar ratio in the Cl•Mg–Al LDH remained at approximately 0.9, even when the pH value after the reaction was high. The diameter of Cl^- (3.62 Å) is larger than that of OH^- (2.80 Å) [31, 42]; therefore, Cl^- has a lower charge density than does OH^-. Thus, we expected a preferential intercalation of OH^- into the Mg–Al LDH interlayer. However, the prepared Cl•Mg–Al LDH also contained Cl^- as the intercalated anion. This suggests that Cl^-, rather than OH^-, reacted preferentially as the "A^{n-}" component, referred to in Eq. (6), with the Mg and Al species in the preparation process of Cl•Mg–Al LDH. During the Cl•Mg–Al LDH preparation, the presence of high Cl^- concentrations prevented the reaction of OH^- with the Mg and Al species, thereby preventing the intercalation of OH^- into the Mg–Al LDH interlayer. At the $[Cl^-]_T$ value of 0.02 M, the Cl/Al molar ratio in the Cl•Mg–Al LDH was approximately 0.9 until the pH value reached 11.5. For pH values higher than 11.5, the Cl/Al molar ratio decreased with an increase in the pH value. With an increasing OH^- concentration, OH^-, rather than Cl^-, preferentially reacted as the "A^{n-}" component, referred to in Eq. (6), with the Mg and Al species and intercalated into the Mg–Al LDH interlayer. At the $[NO_3^-]_T$ value of 0.11 M, the NO_3/Al molar ratio in the NO_3•Mg–Al LDH decreased with increasing pH after the reaction; this contrasted the behavior of the 0.11 M $[Cl^-]_T$. Given that NO_3^- has a larger diameter than Cl^- (3.78 and 3.62 Å, respectively [42]), its interposition between the brucite-like layers would be very little during the Mg–Al LDH formation. Accordingly, OH^-, which has a smaller diameter, reacts more readily with the Mg and Al species, thereby preventing NO_3^- intercalation. As shown in Figure 5, the Cl/Al molar ratio in the Cl•Mg–Al LDH decreased with an increasing pH at low $[Cl^-]_T$ values, suggesting that the anion concentration has a significant effect on its intercalation into the Mg–Al LDH interlayer. The relationships between the Cl^- and NO_3^- concentrations, and the respective molar ratios of Cl/Al and NO_3/Al in the precipitates, are shown in Figure 6. For the Cl•Mg–Al LDH, the Cl/Al molar ratio was approximately 0.5 at the lowest $[Cl^-]_T$. With a slight increase in the $[Cl^-]_T$, the Cl/Al ratio increased to a value greater than 0.8. Despite further increases in the $[Cl^-]_T$, the Cl/Al molar ratio remained between 0.7 and 0.9. For the NO_3•Mg–Al LDH, the NO_3/Al molar ratio was approximately 0.5 at the lowest $[NO_3^-]_T$ value, similar to the molar ratio obtained for the Cl•Mg–Al LDH. With a slight increase in the $[NO_3^-]_T$, the NO_3/Al ratio increased to approximately 0.7.

Further increases in the $[NO_3^-]_T$ did not change the NO_3/Al molar ratio. For molar ratios of around 0.5, at the lowest $[Cl^-]_T$ or $[NO_3^-]_T$, the remaining intercalated anions were accounted for by OH^-. In addition to the hydroxyl groups contained in the Mg–Al LDH host layers, OH^- was found to intercalate to a high degree during the formation of Mg–Al LDHs. The Cl^- and NO_3^- appeared to react easily with the Mg and Al species, and intercalated to high degrees into the Mg–Al LDH interlayer with increasing the $[Cl^-]_T$ and $[NO_3^-]_T$; however, the molar ratio of neither Cl/Al nor NO_3/Al increased to 1.0, even at high $[Cl^-]_T$ or $[NO_3^-]_T$. Thus, the effect of OH^- could not be completely excluded. During the formation of the Mg–Al LDH host layer, some amount of OH^- existed in the vicinity of the layer in solution. This proximity of the formed host layer to OH^- in solution dictates that OH^- would also function as an intercalation anion in all solution conditions.

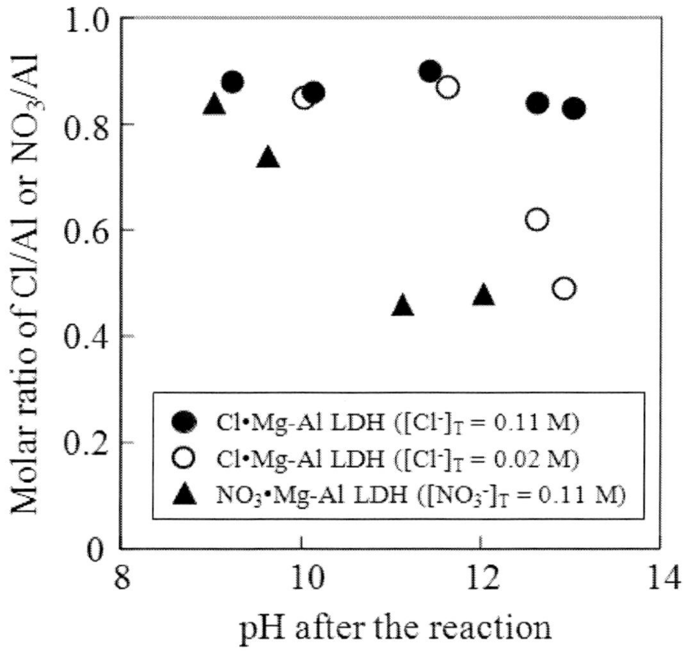

Initial Mg/Al molar ratio in solution was 4.

Figure 5. Effect of pH after the reaction on the molar ratio of Cl/Al or NO_3/Al in the precipitates.

The results shown in Figs. 5 and 6 illustrate that when the solution concentrations of Cl^- and NO_3^- were similar, the Cl/Al molar ratio was greater than the NO_3/Al molar ratio in the precipitate. This indicates that Cl^-, which has a higher charge density than does NO_3^-, preferentially reacted with the Mg and Al species to form a Mg–Al LDH structure. Further, it indicates that intercalation was governed by the charge density of the anion. This conclusion is in good agreement with the anion-exchange property of Mg–Al LDHs and the reconstruction of Mg–Al LDHs from Mg–Al oxide [1, 7, 31, 38]. The solution concentration of Cl^- or NO_3^- controlled the degree of intercalation. Although Cl^- and NO_3^- have lower charge densities than OH^-, they more readily reacted with the Mg and Al species because of their high activity in solution, thereby resulting in the high degree of intercalation demonstrated in this chapter. The relationship between the Mg/Al molar ratio and those of Cl/Al and NO_3/Al in the precipitates shows that the molar ratios of both Cl/Al and NO_3/Al

decreased at a lower Mg/Al molar ratio, owing to the intercalation of OH^-. These results are in agreement with the properties for the reconstruction reaction of Mg–Al LDHs from Mg–Al oxide [7, 38, 39]. For the Mg–Al LDH preparation, larger amounts of negative charges (anions with a high charge density) are required to compensate for an increased percentage of Al^{3+} in a Mg–Al LDH with a lower Mg/Al molar ratio, i.e., with a high value of x in Eq. (6).

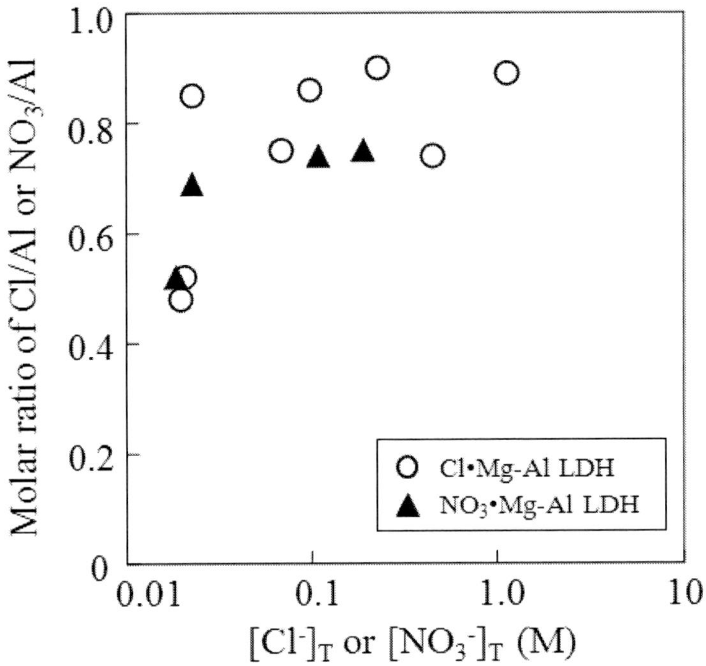

Initial pH value = 10; Initial Mg/Al molar ratio in solution was 4.

Figure 6. Effect of the concentration of Cl^- or NO_3^- on the molar ratio of Cl/Al or NO_3/Al in the precipitates.

Particle Properties Depending on the Preparation Methods

In order to practically utilize Mg–Al LDH for wastewater treatment, the LDH is required to have good solid-liquid separation properties. For the preparation of Mg–Al LDH, the obtained suspensions usually contain very small particles; therefore, the solid-liquid separation process requires a very long time due to its poor sedimentation and filtration properties. Practical wastewater treatment requires large amounts of Mg–Al LDH; accordingly, it is required to prepare Mg–Al LDH with the rapid solid-liquid separation properties. Furthermore, Mg–Al LDH is suspended in wastewater to absorb hazardous materials; hence, another requirement is the rapid filtration of the LDH particles from the wastewater. In order to realize these requirements, it is desirable to prepare Mg–Al LDH with larger particle sizes.

Some researchers have prepared Mg–Al LDHs by homogeneous precipitation using urea hydrolysis and hydrothermal reaction methods in order to control the particle size and morphology [43–46]. A primary aim of these studies is to prepare monodisperse particles with a large primary particle size. For example, Costantino et al. have focused on methods such as precipitation from homogeneous solutions, and found that the hydrolysis of urea in the presence of a mixture of Mg^{2+} and Al^{3+} salts leads to the formation of Mg–Al LDH with a good crystallinity degree and a narrow distribution of particle size [43]. Choy et al. have prepared Mg–Al LDH particles by the hydrothermal aging of the aqueous solutions of Mg^{2+} and Al^{3+} after NaOH titration [44]. From the obtained SEM images, it was noticed that the particle size of the hydrothermally prepared samples increased in proportion to aging time (particle sizes of 85–120 nm over 12–72 hours) and reaction temperatures (particle sizes of 115–340 nm over 100–180°C). Ogawa et al. have prepared a Mg–Al LDH through a homogeneous precipitation method, utilizing urea hydrolysis. An SEM analysis revealed that the products were well-defined Mg–Al LDH particles. Particle sizes of 2 to 20 μm were obtained by carefully controlling the reaction temperatures and the concentrations of the reactants [45]. Using the urea hydrolysis method, Forano et al. have prepared monodisperse sub-micron sized Mg–Al LDH [46]. The as-prepared compounds displayed platelet-like primary particles with a hexagonal shape. However, the hydrothermal method requires high temperatures and pressures, and the urea method produces additional by-products, such as NH^{4+}.

In contrast, our aim was to prepare Mg–Al LDH particles with greater solid-liquid separation properties, through intergrowth, by the aggregates of secondary particles at ordinary temperatures and pressures. In order to do so, we clarified the effects of the preparation methods on the particle properties of prepared Mg–Al LDH, such as the particle size distribution, the particle morphology, and the sedimentation properties. We studied the following three methods of preparing CO_3•Mg–Al LDHs: (i) the addition of a Mg-Al solution to a NaOH-Na_2CO_3 solution, (ii) the addition of a NaOH-Na_2CO_3 solution to a Mg-Al solution, and (iii) the addition of a Mg-Al solution to a Na_2CO_3 solution at constant pH [47, 48]. We found that the ideal preparation method was (iii)—the addition of a Mg-Al solution to a Na_2CO_3 solution at a constant pH value adjusted by a NaOH solution. This preparation method resulted in CO_3•Mg–Al LDH particles with uniform primary particle sizes, good sedimentation properties, and narrow and manageable distributions of the aggregates of secondary particles. A future area of study is the refinement of the abovementioned method (iii) in the industrial scale preparation of Mg–Al LDH.

CONCLUSION

In conclusion, we summarize the results of our various methods of preparing Mg–Al LDH.

In the soda ash process using seawater, Na_2CO_3 solution was added to the Ca^{2+}-free artificial seawater containing $AlCl_3$, until a pH of 10 was obtained, and it was stirred at 60°C for 1 hour. A CO_3•Mg–Al LDH sample was quantitatively precipitated without by-products for initial Mg/Al molar ratios of 2.0–2.3.

In the milk of lime process using seawater, $Ca(OH)_2$ slurry was added to artificial seawater containing $AlCl_3$, until a pH of 10.5 was obtained, and stirred at 60°C for 1 hour. A Mg–Al LDH sample was precipitated without by-products with initial Mg/Al molar ratios of 2.0–4.0. The initial Mg/Al molar ratios (for which each recovery of Mg^{2+} and Al^{3+} from the solution was quantitatively) were 2.4–3.0. Mg–Al LDH was precipitated from the solution at 25°C for an initial Mg/Al molar ratio of 3.0 when 0.7–10.0 wt% $Ca(OH)_2$ slurry and $Ca(OH)_2$ powder were used. SO_4^{2-} and Cl^- were intercalated in the interlayer of Mg–Al LDH prepared from natural seawater.

In the dolomite process, Mg–Al LDH was obtained as a single phase at 25°C by the addition of calcined dolomite to mixtures of $MgCl_2$ and $AlCl_3$ at $((Mg)_D + (Mg)_{Cl})$/Al molar ratios of 2.0 and 3.0. Mg–Al LDH could also be prepared at an ambient temperature using a mixture of PAC (a new aluminum source) and calcined dolomite. We speculated that Mg–Al LDH is formed through the reactions of Mg^{2+}, $Al(OH)_4^-$, and anions; these reactions are expressed by Eq. (6). A Mg-Al solution, together with an alkali (used as pH control), is added dropwise to deionized water. In the local environment where the droplets of alkali are added, the pH is very high. This high pH results in the formation of $Al(OH)_4^-$ from $Al(OH)_3$; this $Al(OH)_3$ forms in solution with the addition of the Mg–Al solution. The $Al(OH)_4^-$ quickly reacts with Mg^{2+} and anions, leading to the formation of Mg–Al LDH. We confirmed that the characteristics of anion intercalation into the interlayer of Mg–Al LDH during preparation are similar to those of the anion-exchange reactions of Mg–Al LDHs, as well as the reconstruction reactions of Mg–Al LDHs from Mg-Al oxides. We demonstrated the following results: (1) Cl^-, which has a higher charge density than NO_3^-, more easily reacted with the Mg and Al species to form the Mg–Al LDH structure, thereby resulting in a greater intercalation of Cl^- into the Mg–Al LDH interlayer; and (2) for Mg–Al LDHs with lower Mg/Al molar ratios, OH^- (which has a higher charge density than Cl^- and NO_3^-) was more likely to interact with the Mg and Al species to form the Mg–Al LDH structure, thereby blocking the intercalation of Cl^- and NO_3^-. Furthermore, we showed that high concentrations of Cl^- and NO_3^- in solution regulated their intercalation into the Mg–Al LDH interlayer. The high activity of Cl^- and NO_3^- in solution facilitated the anion reactions with Mg and Al species to form Mg–Al LDHs, thereby resulting in a high degree of anion intercalation into the interlayer of Mg–Al LDHs. The ideal preparation method of Mg–Al LDH, at ordinary temperatures and pressures, is considered to be the addition of an Mg–Al solution to a Na_2CO_3 solution at a constant pH, adjusted by a NaOH solution. This preparation method resulted in CO_3•Mg–Al LDH particles with uniformly sized primary particles, good sedimentation properties, and the narrow and manageable distributions of the aggregates of secondary particles.

REFERENCES

[1] Miyata S. *Clay. Clay Miner.* 1983, *31*, 305.

[2] Cavani F.; Trifiro F.; Vaccari A. *Catal. Today*, 1991, *11*, 173.

[3] Ingram L.; Taylor H.F.W. *Mineral. Mag.* 1967, *36*, 465.

[4] Allmann R. *Acta Crystallogr.* 1968, *B24*, 972.

[5] Kameda T.; Miyano Y.; Yoshioka T.; Uchida M.; Okuwaki A. *Chem. Lett.* 2000, *29*, 1136.

[6] Kameda T.; Yoshioka T.; Uchida M.; Miyano Y.; Okuwaki A. *Bull. Chem. Soc. Jpn.* 2002, *75*, 595.

[7] Kameda T.; Yabuuchi F.; Yoshioka T.; Uchida M.; Okuwaki A. *Water Res.* 2003, *37*, 1545.
[8] Kameda T.; Yoshioka T.; Mitsuhashi T.; Uchida M.; Okuwaki A. *Wat. Res.* 2003, *37*, 4045.
[9] Kameda T.; Yoshioka T.; Hoshi T.; Uchida M.; Okuwaki A. *Sep. Purif. Technol.* 2005, *42*, 25.
[10] Kameda T.; Yoshioka T.; Hoshi T.; Uchida M.; Okuwaki A. *Sep. Purif. Technol.* 2006, *51*, 272.
[11] Yoshioka T.; Kameda T.; Miyahara M.; Uchida M.; Mizoguchi T.; Okuwaki A. *Fresenius Environ. Bull.* 2007, *16*, 928.
[12] Yoshioka T.; Kameda T.; Miyahara M.; Uchida M.; Mizoguchi T.; Okuwaki A. *Chemosphere* 2007, *69*, 832.
[13] Kameda T.; Saito M.; Umetsu Y. *J. Alloys Compd.* 2005, *402*, 46.
[14] Kameda T.; Saito S.; Umetsu Y. *Sep. Purif. Technol.* 2005, *47*, 20.
[15] Kameda T.; Saito M.; Umetsu Y. *Mater. Trans.* 2007, *48*, 2225.
[16] Kameda T.; Takeuchi H.; Yoshioka T. *Sep. Purif. Technol.* 2008, *62*, 330.
[17] Kameda T.; Yamazaki T.; Yoshioka T. *Chem. Lett.* 2009, *38*, 522.
[18] Kameda T.; Saito M.; Umetsu Y. *Mater. Trans.* 2009, *50*, 1541.
[19] Kameda T.; Yamazaki T.; Yoshioka T. *Bull. Chem. Soc. Jpn.* 2009, *82*, 1436.
[20] Kameda T.; Tsuchiya Y.; Yamazaki T.; Yoshioka T. *Solid State Sci.* 2009, *11*, 2060.
[21] Kameda T.; Yamazaki T.; Yoshioka T. *Solid State Sci.* 2010, *12*, 946.
[22] Kumura T.; Imataki N.; Hasui K.; Inoue T.; Yasutomi K. Jp. Patent 46/2280 (1971).
[23] Ross G.J.; Kodama H. *Amer. Mineral.* 1967, *52*, 1036.
[24] Miyata S. *Clays Clay Miner.* 1975, *23*, 369.
[25] Miyata S. *Clays Clay Miner.* 1977, *25*, 14.
[26] Mirsa C. U.S. Patent 4 904 457, (1990).
[27] Tsuboi T.; Midera T.; Kabeya R. Jpn. Patent 6/329410, (1994).
[28] Kasai T.; Mizota T.; *Shigen-to-Sozai* 1996, *112*, 131.
[29] Kameda T.; Yoshioka T.; Uchida M.; Okuwaki A. *Mol. Cryst. Liq. Cryst.* 2000, *341*, 407.
[30] Ishizdka S.; Terada M. *Tokyo Kogyo Shikensho Hokoku* 1957, *52*, 201.
[31] Sato T.; Wakabayashi T.; Shimada M. *Ind. Eng. Chem. Prod. Res. Dev.* 1986, *25*, 89.
[32] Kameda T.; Yoshioka T.; Uchida M.; Okuwaki A. *Phosphorus, Sulfur and Silicon* 2002, *177*, 1503.
[33] Kameda T.; Yoshioka T.; Uchida M.; Okuwaki A. *J. Mater. Sci. Lett.* 2002, *21*, 1747.
[34] Vanderlaan R.K.; White J.L.; Hem S.L. *J. Pharm. Sci.* 1982, *71*, 780.
[35] Kameda T.; Yoshioka T.; Yabuuchi F.; Uchida M.; Okuwaki A. *J. Mater. Sci.* 2007, *42*, 2194.
[36] Pourbaix M. Atlas of Electrochemical Equilibria in Aqueous Solutions. National Association of Corrosion Engineers, Houston (1974).
[37] Kameda T.; Yoshioka T.; Yabuuchi F.; Uchida M.; Okuwaki A. *Bull. Mater. Sci.* 2008, *31*, 625.
[38] Sato T.; Onai S.; Yoshioka T.; Okuwaki A. *J. Chem. Tech. Biotechnol.* 1993, *57*, 137.
[39] Sato T.; Kato K.; Yoshioka T.; Okuwaki A. *J. Chem. Tech. Biotechnol.* 1992, *55*, 385.
[40] Sato T.; Tezuka M.; Endo T.; Shimada M. *React. Solids* 1987, *3*, 287.

[41] Lukashin A.V.; Kalinin S.V.; Nikiforov M.P.; Zhirov A.I.; Kheifets L.I.; Privalov V.I.; Vertegel A.A. *Mat. Res. Soc. Symp. Proc.* 1999, *547*, 239.
[42] Kapustinskii A.F. *Quart. Rev.* 1956, *283.*
[43] Constantino U.; Marmottini F.; Nocchetti M.; Vivani R. *Eur. J. Inorg. Chem.* 1998, *1998*, 1439.
[44] Oh J.M.; Hwang S.H.; Choy J.H. *Solid State Ion.* 2002, *151*, 285.
[45] Ogawa M.; Kaiho H. *Langmuir* 2002, *18*, 4240.
[46] Pagano M.A.; Forano C.; Besse J.P. *J. Mater. Chem.* 2003, *13*, 1988.
[47] Kameda T.; Umetsu Y. *Proc. MMIJ Faul Meeeting* 2003, *Vol.(C)(D)*, 245.
[48] Kameda T.; Nakayama Y.; Umetsu Y. *Proc. MMIJ Annual Meeting* 2004, *Vol.(II)*, 103.

In: Precipitation: Prediction, Formation, Environmental Impact ISBN: 978-1-62100-447-9
Editors: Henry Dohring, Jeremy Dixon

Chapter 5

MAIN WATER-SOLUBLE IONS IN PRECIPITATION OF THE CENTRAL MEDITERRANEAN REGION

Dragana S. Đorđević*[*], Pavle Đurašković[2], Ivana Tošić[3], Miroslava Unkašević[3] and Ljubiša Ignjatović[4]
[1]Institute of Chemistry, Technology and Metallurgy – Chemistry Center, University of Belgrade, Belgrade, Serbia
[2]Hydrological and Meteorological Service of Montenegro, Podgorica, Montenegro
[3]Department of Meteorology, University of Belgrade, Belgrade, Serbia
[4]Faculty of Physical Chemistry, University of Belgrade, Belgrade, Serbia

ABSTRACT

The chemical content of atmospheric deposition is the signature of several interacting physical and chemical mechanisms: i.e., emission and source amplitude; transport processes and dynamics of the atmosphere; atmospheric chemical reactions; and removal processes, among others. Studying deposition processes and amounts thus allow for tracing the spatio-temporal evolution of atmospheric chemistry. It is also an important way for distinguishing between natural and anthropogenic influences. Sulfate, nitrate and ammonium ions are formed in the atmosphere in gas-particle conversion processes. Sulfates represent one of the main components of atmospheric aerosols. In southern Europe, rainwater has similar concentrations of acidic anions to those of other parts of Europe, but the average precipitation is not acid due to neutralizing role of basic cations. The Mediterranean Sea is enclosed basin, which is susceptible to dust transport from desert sources in North Africa (Sahara) and the Middle East (the Arabian Peninsula and Syria), lying on its periphery. The climatic conditions are favorable for the transport of dust from North Africa into the eastern Mediterranean atmosphere during the spring, while other conditions favor transport into the western Mediterranean atmosphere during the summertime. An investigation of wet deposition of water soluble Na^+, K^+, Mg^{2+}, Ca^{2+}, Cl^-, SO_4^{2-}, NO_3^- and NH_4^+ were done in south-eastern Adriatic coast in the period from

January 1995 until December 2000. The highest precipitation and the highest frequencies of air masses coming from the western Mediterranean cause the highest Wet Deposition (WD) of the main water soluble ions. The most abundant WD of SO_4^{2-} in investigated region was caused by air masses coming from the western Mediterranean and western and central Europe. The sea salt components (Na^+, Cl^-, Mg^{2+}) are significantly correlated. The correlations between Ca^{2+} and Mg^{2+}, Ca^{2+} and SO_4^{2-} and Ca^{2+} and K^+ indicate that the main terrigeneous ions were associated with formation of eolian particles which include gypsum ($CaSO_4$), calcite ($CaCO_3$) and dolomite ($CaMg(CO_3)_2$). Although the Volume Weighted Mean (VWM) of Na^+, K^+, Mg^{2+}, Ca^{2+}, Cl^-, SO_4^{2-}, NO_3^- and NH_4^+ that are coming in air masses from the continental parts of Europe are significant, the total Wet Deposition (WD) is dominated by precipitation coming over Mediterranean due to abundances of precipitation from this segment.

1. INTRODUCTION

Wet- and dry- deposition of chemical species onto the earth's surface play a key role in controlling the concentration of gases and aerosols in the troposphere. The chemical content of atmospheric deposition is the signature of several interacting physical and chemical mechanisms: i.e., emission and source amplitude; transport processes and dynamics of the atmosphere; atmospheric chemical reactions; and removal processes, among others. Studying deposition processes and amounts thus allow for tracing the spatio-temporal evolution of atmospheric chemistry. It is also an important way for distinguishing between natural and anthropogenic influences. In regions where natural biogeochemical cycles are perturbed by human activities, atmospheric deposition can either be important sources of toxic substances or of nutrients for the ecosystems [1-2].

Mineral dust and sea salt are major components of particulate matter in atmosphere. Desert dust accounts for more than 50% of the global aerosol load [3-5] and the long range transport dust particles can influence the composition and dynamic state of the atmosphere thousands of kilometers downwind of their source region [6].

Sulphate, nitrate and ammonium ions are formed in the atmosphere in gas-particle conversion processes. Sulphates represent one of the main components of atmospheric aerosols. They are mainly obtained by the oxidation of SO_2, which was either directly emitted to the atmosphere (burning of fossil fuels, industrial processes, volcanoes, combustion of biomasses) or by the oxidation of lower oxidation state sulphur compounds, mainly DMS, which is the dominant source of SO_2 in marine atmospheres. The main sources of NO_x are combustion of fossil fuels in high temperature processes (traffic, power plants, industry, and domestic fire-boxes), soil (microbiological activity), combustion of biomass, lightening, *etc.* [7]. The vapour pressure ensures that H_2SO_4 and HNO_3 are in the particulate phase in atmospheric conditions (with exception of HNO_3 in warm condition). Sulphate and nitrate aerosols can also be obtained by reaction of HNO_3 and H_2SO_4 in presence of salts or ammonia results in the neutralisation of the acids and, together with NaCl from marine aerosol [7-8]. The sulphates in the atmosphere can also originate from other sources: *e.g.*, primary marine aerosol, gypsum $CaSO_4$ from (resuspended) dust from the Sahara [9-10] *etc.*

The ammonia in precipitation is originating from the particulate phase. Seinfild (1986) reported that generally in the atmosphere ammonia occurs as $(NH_4)_2SO_4$ but it is also possible to found in the form NH_4NO_3 [11].

The residence time of SO_2 and NO_x in the troposphere is 1 – 3 days [10], while the residence time of sulphates and nitrates is somewhat longer: for nitrates it is 3-9 days [7], while it can even be 10 days for sulphates during dry periods in the Mediterranean region [12].

Wet deposition refers to the natural processes by which material is scavenged by atmospheric hydrometeors (cloud and fog drops, rain and snow) and is consequently delivered to the Earth's surface. Washout usually refers to in-cloud scavenging and rainout to below-cloud scavenging by falling rain, snow and so on.

1) Precipitation scavenging that is removal of species by rain cloud.
2) Cloud interception, the impaction on cloud droplets on the terrain usually at the top of tall mountains.
3) Fog deposition that is removal of material by settling fog droplets.
4) Snow deposition removal of material during a snowstorm.

In all of the above processes three steps are necessary for wet removal of material. Specifically, the species (gas or aerosol) must first be brought into the presence of condensed water. Then, the species must be scavenged by the hydrometeors, and finally it needs to be delivered to the Earth's surface [7]. Wet removal pathways depend on multiple and composite processes, involve numerous physical phases, and are influenced by phenomena on a variety of physical scales. The challenge of understanding processes that operate on the microscale (10^{-6} m) and the macroscale (10^{6} m) makes wet deposition one of the most complex atmospheric processes [7]. The complexity of the wet removal process led early investigators to attempt to quantify the relationship between airborne species concentrations, meteorological conditions, and wet deposition rates by lumping the effects of these processes into a few parameters. The rate of transfer of a soluble gas or a particle into rain droplets below a cloud can be approximated by first-order relationships:

$$W^{i}_{gas/rain} = \Lambda_{ig} C_{i.gas}$$
$$W^{i}_{aerosol,rain} = \Lambda_{ip} C_{i.part}$$

where Λ_{ig} and Λ_{ip} are the scavenging coefficients for species i in the gas and the particulate phases, respectively. Scavening coefficients in general are a function of location, time, rainstorm characteristics, and the aerosol size distribution of species i [7].

Data from North American Deposition Program (NADP), EMEP network, IGAC DEBITS Africa (IDAF), Acid Deposition Monitoring Network in East Asia (EANET), Integrated Monitoring Program on Acidification of Terrestrial Ecosystems (IMPACT) in China etc., are used for investigation large spatial and temporal heterogeneity of precipitation and its chemical composition. Two major local wet deposition maxima of nitrate are in China resulting mainly from the strong emissions of nitrate oxides (NO_x) mostly from fossil energy use, and in Central Africa, where anthropogenic emissions are dominated by residential biofuel use and biomass burning. Furthermore, large deposition rates are also calculated for the eastern USA and Western Europe, both mainly from fossil fuel related NO_x emission sources. In Southeast Asia, Amazonia and central Africa additionally the natural NO_x emissions from soils significantly contribute to the nitrate contents of precipitation. NOx production from lightning plays a minor role for wet deposition (globally ≈ 5% of the total emissions are from lightning, mainly over the land masses in the tropics). Over the tropical continents the large amounts of precipitation lead to an almost complete scavenging of nitric acid (HNO_3) from the gas phase, while in the northern hemispheric storms tracks highly efficient wet deposition of HNO_3 is over the Atlantic Ocean. The latter is also a consequence of the incomplete scavenging close to the North America east coast (due to lower and more episodic characteristics of the precipitation compared to the tropics) and subsequent westward transport before the atmospheric pollution is removed from the atmosphere.

The wet deposition of sulfate occurs mainly in the vicinity of the emission sources of sulfur dioxide (SO_2), of which a small fraction is oxidized in the gas phase, whereas the more important oxidation pathway is in the aqueous phase. The largest wet deposition fluxes are calculated for China, where the SO_2 emissions are strongest. Additional local maxima are computed for the industrialized regions in the eastern USA and central Europe. In many of these regions energy production relies on coal types with higher sulfur content, resulting in substantial SO2 emissions and consequent sulfate deposition.

The wet deposition fluxes of NH_x compounds, are also highest in regions where the emissions are typically strongest, i.e., in China and India, and slightly lower in Western Europe, eastern USA, Central Africa and the northern part of South America. While in East Asia, North America and Europe both agricultural and industrial emissions are the main sources, in Africa and South America biofuel use and biomass burning dominate the atmospheric ammonia burden. In contrast to nitrate and sulfate higher wet deposition occurs in the tropical, both over land and ocean. This partly results from a neglect of the chemical loss of NH_3 in the gas phase (oxidation by OH). Furthermore, the effects of uncertainties of the simulated precipitation rates become increasingly important for moderately soluble species such as NH_3. However, since observations are very scarce over the tropical oceans, it is difficult to judge if these features are real or an artifact of the model [13].

Process understanding scavenging of gases and particles by precipitation has continued to advance, with important developments during the last decade. The applications of wet deposition schemes are very important in the Long-Range Transport (LRT) models and increasingly in global chemistry-transport models (CTM) [14]. These two applications make very different demands on available knowledge and understanding. In the case of LRT models in Europe (e.g. EMEP), the applications from part of the integrated assessment process and within the user community the pressure to provide ever finer spatial scale estimates of inputs presents challenges in the capability of LRT models and the meteorological models on which they depend [15]. Wet deposition of acidifying compounds increased during the 1960s and 1970s following decades of increasing emissions of SO_2 and NO_x in industrialized countries. The emission of SO_2 has decreased since the 1980s in most northern countries. In particular, Spanish SO_2 emissions have been reduced by 47% between 1980 and 1996 [16]. The effect of these changes on precipitation chemistry and on the deposition fluxes of S, N, and acidity is of great interest because of the need to evaluate the environmental consequences of the abatement strategies. In northern Europe, decreases in rainwater sulfate and H^+ concentrations have been related to declining SO_2 emissions. The decries of acid deposition over the last two decades seems to have prompted a recovery of acidified soils and aquatic ecosystems across many regions of Europe and North America. In southern Europe, rainwater has similar concentrations of acidic anions to those of other parts of Europe, but the average precipitation is not acid due to neutralizing role of basic cations [17].

The Mediterranean Sea is enclosed basin, which is susceptible to dust transport from desert sources in North Africa (Sahara) and the Middle East (the Arabian Peninsula and Syria), lying on its periphery [18-21]. The climatic conditions are favourable for the transport of dust from North Africa into the eastern Mediterranean atmosphere during the spring, while other conditions favour transport into the western Mediterranean atmosphere during the summer-time [18].

The Hybrid Single-Particle Lagrangian Integrated Trajectory (HYSPLIT) model, [22-24] is a commonly used air modelling program that can calculate air mass paths from one region to another [25]. The HYSPLIT model is a complete system that is designed to support a wide range of simulations related to regional or long-range transport, dispersion, and deposition of air pollutants. The trajectory calculation was achieved by the time integration of the position of an air parcel as it is transported by 3-D winds [23].

2. EXPERIMENTAL

A bottle and funnel with a small net between them, the function of which was to prevent contamination of the samples (leaves, small branches, insects, *etc.*) was used for sampling. Samples of the precipitation were collected daily from 8 am to 8 am (meteorological time) the following day in the period from January 1995 until December 2000 only in rainy days. The collection surface was continuously exposed during 24 hours without coverage of the precipitation samples, thus samples were included also part of dry deposition. The samplers were at a height 1.5 m above the ground. Before use, the sample bottles and plastic equipment (PE bottles and funnels) were submerged in water for 24 h, washed with detergent in warm water to mechanically remove any dirt and then rinsed with deionised water until the electrical conductivity of the washings was less than 2 μScm^{-1} [26-27]. The samples are stored in a deep freeze environment, thus avoiding microbial activities and possible alteration of chemical composition. The amounts of anions in the precipitation were determined spectrophotometrically as follows: Cl^- by the mercury thiocyanate-iron method, NO_3^- by the Griess method and SO_4^{2-} by the barium perchlorate-Thorin method. The amount of NH_4^+ was determined spectrophotometrically by the Indophenol Blue method. The amounts of Na^+ and K^+ in the precipitation was determined by the FAES method and the amounts of Mg^{2+} and Ca^{2+} by the FAAS method.

Quality control of results was provided by direct comparison with the analytical laboratories which indicated that laboratory uncertainties did not significantly contribute to the differences in the networks' data. During investigation period our laboratory participated in EMEP [28] and WMO-GAV laboratory intercomparasions [29] of major ions concentrations (Na^+, Mg^{2+}, K^+, Ca^{2+}, NH_4^+, SO_4^{2-}, NO_3^- and Cl^-).

The air mass origin for each sampling day was studied by air back-trajectories. The following origins of the air masses were considered: northern Europe (NE), eastern Europe – north-eastern Europe (EE-NE); eastern Mediterranean – southeastern Europe (EM-SE); Africa-Central Mediterranean (A-CM); western Mediterranean (WM); western Europe – Central Europe (WE-CE) [27].

3. Results and Discussion

The sampling station is located at the entry of the Boka Kotorska Gulf (Figure 1) in the eastern suburban of the town Herceg-Novi ($18^{0}33'$ N, $42^{0}27'$ E).

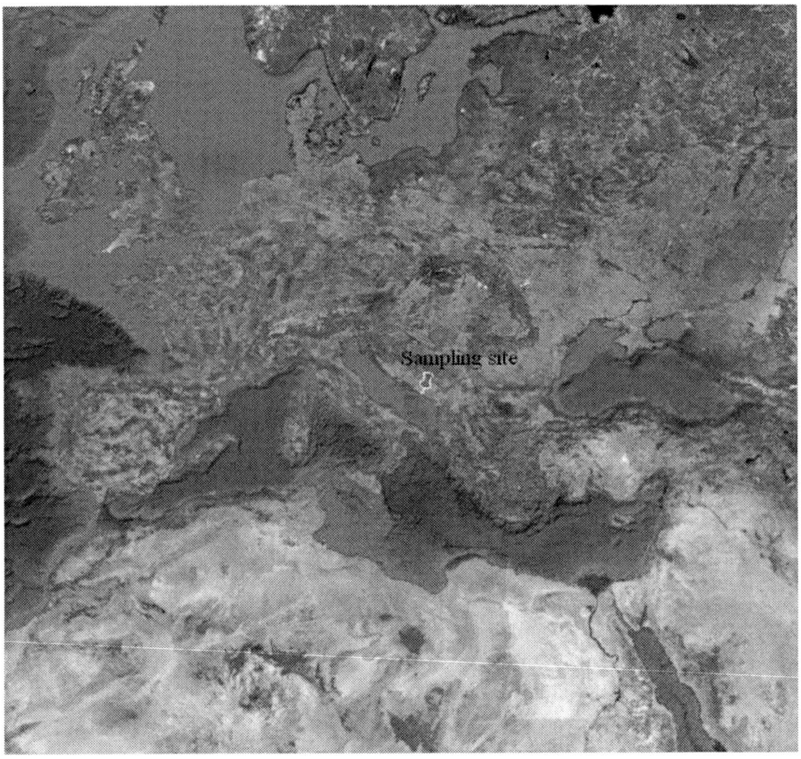

Figure 1. Sampling location.

3.1. Ion Concentrations and Wet Depositions

The Volume-Weighted Mean (VWM) concentrations of the ionic constituents measured in rainwater in south-east Adriatic coast region were calculated using the following equation:

$$VWM(\mu eqL^{-1}) = \frac{\sum_{i=1}^{n=604} C_i P_i}{\sum_{i=1}^{n=604} P_i} \qquad (1)$$

where C_i is the ion concentration for each species in μeq l^{-1} and P_i the precipitation amount for each rainy event in mm. The Wet Deposition (WD) was calculated using the equation:

$$WD = VWM \times P \qquad (2)$$

with VWM being the Volume-Weighted Mean concentration for each species and P the mean rainfall amount measured during the studied period [11].

The calculated Volume Weighted Mean (VWM) and total Wet Deposition (WD) over the whole period of investigation, and according to the air mass origin are shown (Figure 2). There are no significant differences in trends of VWM and WD considering in relation to period (Figure 2a$_1$ and a$_2$) but, observing the directions of air masses from which they come there are significant difference between VWM and WD for air masses coming from Eastern Europe where is found the highest ion concentrations and Volume Weighted Mean of SO_4^{2-} in the precipitation, indicating anthropogenic sources that are consuming coal (coal fired power plants) and liquid fuels with a high content of sulfur and, old technologies without emission control that are still in use in this region. But, due to low the frequency of rainfall and low of their heights WD of ions in precipitations coming from Eastern Europe region is not significant (Figure 2b$_1$ and 2$_2$).

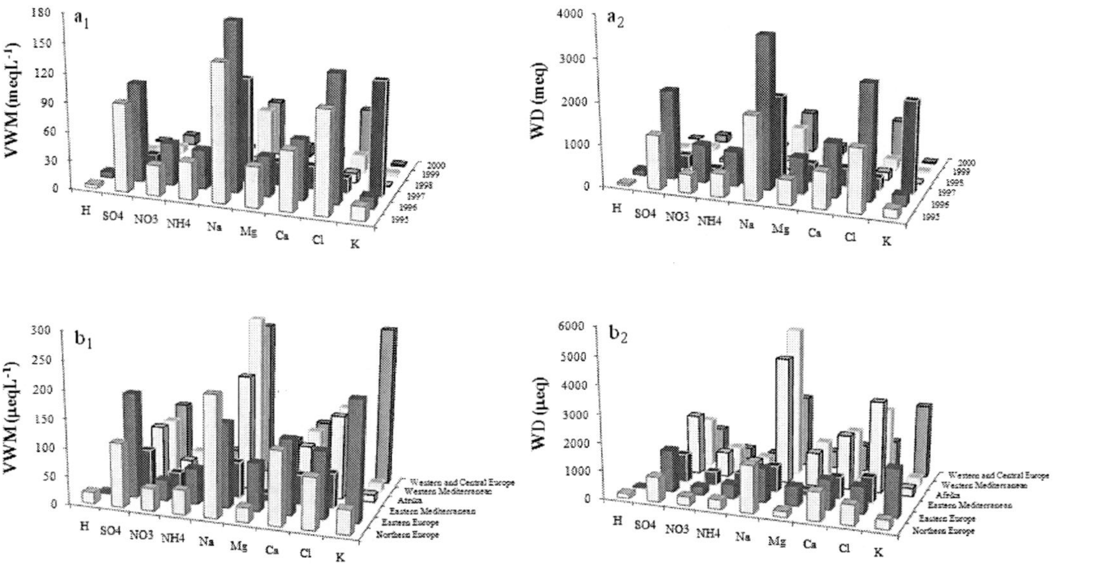

Figure 2. VWM (a) and WD (b) yearly (a$_1$ and a$_2$) and in dominant air masses (b$_1$ and b$_2$).

3.2. Main Correlations

The Spearman coefficients of correlations were calculated for sea salt (Na^+, Cl^-, Mg^{2+} (Figure3a)) and no sea salt components (SO_4^{2-}, NH_4^+, Ca^{2+}, Mg^{2+} and K^+ (Figure3a and b)) in total data set.

The data which refers to sea salt components is accompanied by strong coefficients of correlation (at the confidence level of > 99.9% ($p = 0.000$) between Na^+ and Cl^- ($r = 0.753$); Na^+ and Mg^{2+} ($r = 0.784$); Cl^- and Mg^{2+} ($r = 0.761$). Coefficients of correlation (at the confidence level of > 99.9% ($p = 0.000$) for no sea salt components between SO_4^{2-} and NH_4^+ is $r = 0.458$; SO_4^{2-} and Ca^{2+} ($r = 0.582$) (Figure 3b) and, between Mg^{2+} and Ca^{2+} ($r = 0.675$) and Mg^{2+} and K^+ ($r = 0.666$) (Figure 3c); Ca^{2+} and K^+ ($r = 0.642$).

The sea salt components (Na^+, Cl^-, Mg^{2+}) are significantly correlated. The calcium also derived from soil dust, is correlated with magnesium (which is mainly supplied by sea salt) in air masses coming over sea, indicating jointly transport dust and marine aerosol over

Mediterranean. The significant correlation of Ca^{2+} and SO_4^{2-} is pointing out Sahara dust as origin of $CaSO_4$.

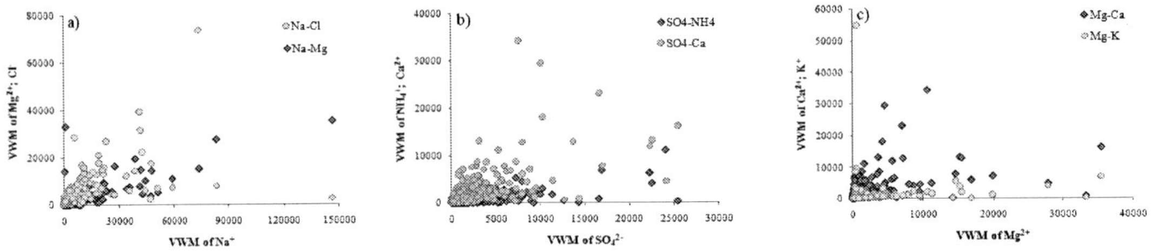

Figure 3. Correlations among sea salt (a), non-sea salt (b) and terrigeneous (c) components.

The analyses of correlations give significant and high coefficients of correlation between Ca^{2+} and Mg^{2+}, Ca^{2+} and SO_4^{2-} and Ca^{2+} and K^+. These correlations indicate that the main terrigeneous ions were associated with formation of eolian particles which include gypsum ($CaSO_4$), calcite ($CaCO_3$) and dolomite ($CaMg(CO_3)_2$). The origin of potassium ion in rainwater is from washout of desert aerosol (from soil wind erosion), aerosols from combustion and biogenic aerosols [11].

3.3. Contents of Non-Sea Salt Fraction Of SO_4^{2-} in Precipitations of Investigated Region

The non-sea salt fraction of SO_4^{2-} ($nssSO_4^{2-}$) contribution is usually calculated using total Na as sea-spray marker. Here, for better accuracy, $nssSO_4^{2-}$ was evaluated from the $ssNa^+$ fraction according to the equation:

$$nssSO_4^{2-} = SO_4^{2-} - 0.253ssNa^+,$$

where $SO_4^{\ 2-}$ is the total sulphate concentration and $ssNa^+$ is the Na^+ actually derived from sea spray. Since some Na^+ derives from continental dust, $ssNa^+$ was calculated using the four-equation system reported below and knowing total Na^+, total Ca^{2+}, the mean Ca^{2+}/Na^+ ratio in the crust ($(Na^+/Ca^{2+})crust$ = 1.78 w/w; [30] and the mean Ca^{2+}/Na^+ ratio in sea water ($(Ca^{2+}/Na^+)seawater$ = 0.038 w/w; [30].

$$ssNa^+ = Na^+ - nssNa^+$$

$$nssNa^+ = nssCa^{2+}(Na^+/Ca^{2+})crust$$

$$nssCa^{2+} = Ca^{2+} - ssCa^{2+}$$
$$ssCa^{2+} = ssNa^+ (Ca^{2+} /Na^+)seawater$$

The non-sea salt fraction of SO_4^{2-} was calculated from the measured Na^+ concentrations and a standard sea water and crust compositions.

Table 1. Non-sea salt fraction of SO_4^{2-} contribution in precipitations from air masses

nssSO_4^{2-} (mg L^{-1}); average ± st. dev (σ)					
NE	EE-NE	EM-SE	A-CM	WM	WE-CE
5.79 ± 4.72	9.37 ± 9.52	5.02 ± 4.73	6.47 ± 6.87	6.44 ± 9.93	6.52 ± 7.44

Figure 4 presents time series of non-sea salt fraction of SO_4^{2-} (nssSO_4^{2-}) in precipitations according to the air mass origin studied by air back-trajectories.

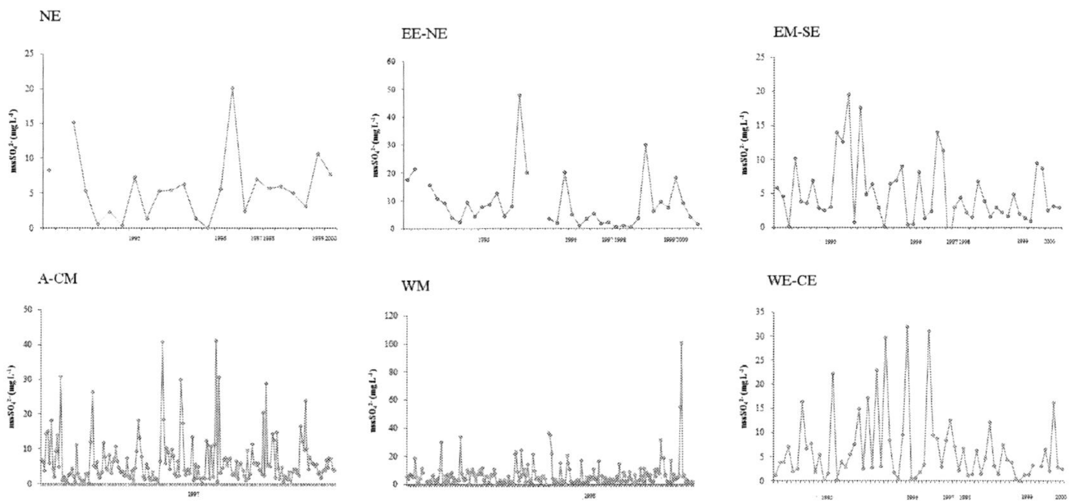

Figure 4. Time series of non-sea salt fraction of SO_4^{2-} in precipitations of dominant air masses.

The highest values of non-sea salt fraction of SO_4^{2-} contribution were found in precipitations of air masses coming from Eastern Europe. Nevertheless, the most significant WD of SO_4^{2-} in the investigated region was caused by air masses coming from Africa ($CaSO_4$ in Sahara's dust) and Central and Western Mediterranean (Figure 2). In the region of the Central Mediterranean, there are several volcanoes, located in South Italy: Vesuvius, Etna and the permanent volcano Stromboli that are continuously emitting sulfur oxides as a natural emission sources.

Figure 5 gives information about rain water pH values for the investigated trajectories groups. The most frequently pH values were around 6.5, for each trajectory group. The higher pH values were noticed in rainfall trajectory group from open sea.

3.4. Sea Salt

In all trajectory groups Na^+ and Cl^- were the most abundant components. These the major sea salt components were also strongly correlated (Figure 3; Figure 6a). The average ion concentration Cl^-/Na^+ ratio of 0.9015 was found and most frequently ratios are about 1 (Figure 6c). On the other hand Na^+/Cl^- average ratio was 1.8909 which is in well accordance with ratio of 1.8 reported for sea water [7].

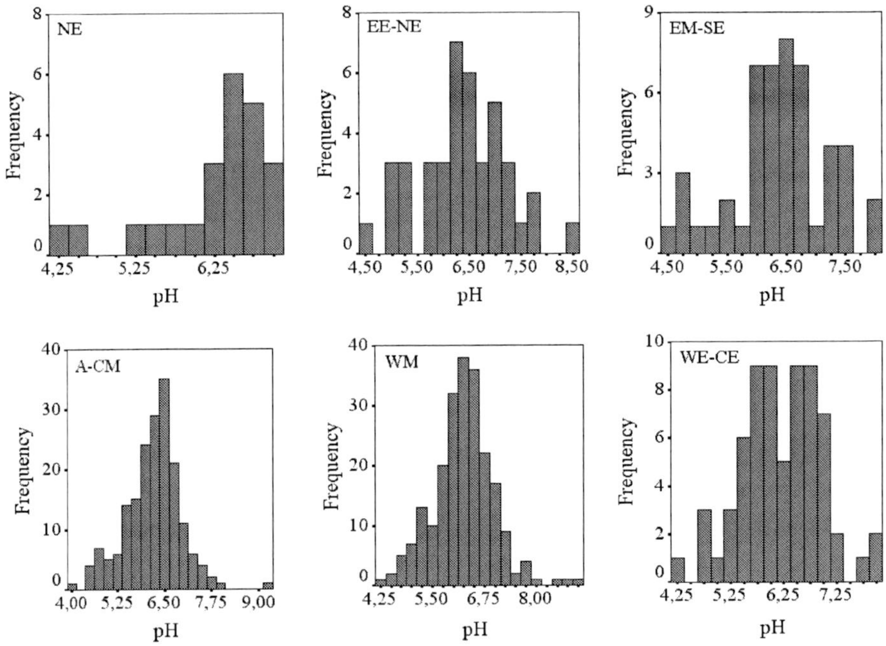

Figure 5. Frequency distributions of pH in dominant trajectory groups.

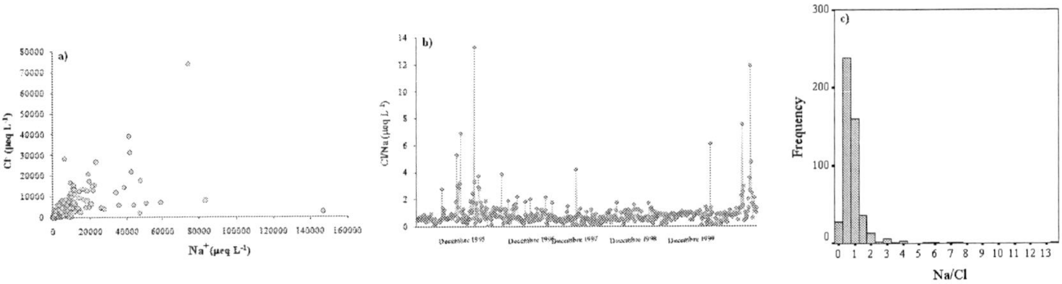

Figure 6. Correlation between Na^+ and Cl^- (a), time serie of Na^+/Cl^- (b) and Na^+/Cl^- distribution (c).

Conclusion

An investigation of wet deposition of water soluble Na^+, K^+, Mg^{2+}, Ca^{2+}, Cl^-, SO_4^{2-}, NO_3^- and NH_4^+ were done at the coast of south-eastern Adriatic sea (Herceg Novi in Montenegro) in the central Mediterranean region.

The sea salt components (Na^+, Cl^-, Mg^{2+}) are significantly correlated. The most abundant Wet Deposition (WD) was caused by air masses coming from the western Mediterranean and western and Central Europe. The highest precipitation and the highest frequencies of air masses coming from Western Mediterranean are causing the highest WD of SO_4^{2-} in relation to air masses coming from continental regions where ion concentration is maximal.

The absolutely highest values of $nssSO_4^{2-}$ were found in precipitations of air mass origin from North Eastern Europe, studied by air back-trajectories.

The average ratio of Na^{+}/Cl^{-} was 1.8909 which is in well accordance with ratio of 1.8 reported for sea water.

REFERENCES

[1] Brimblecombe, P., Acid Rain – Deposition to Recovery, springer edition, repreinted from water, Air Soil Poll. *Focus*, 7 (1–3) (2007) 1-2

[2] Galy-Lacaux C., Laouali D., Descroix L., Gobron N., and Liousse C., Long term precipitation chemistry and wet deposition in a remote dry savanna site in Africa (Niger), *Atmospheric Chemistry and Physics* 9 (2009) 1579-1595.

[3] Andreae, M. O., Charlson, R. J., Bruynseels, F., Storms, H., van Grieken, R., and Maenhaut, W.: Internal mixtures of sea salt, silicates and excess sulfate in marine aerosols, *Science*, 232, 1620–1623, 1986.

[4] Zender, C. S., Miller, R., and Tegen, I.: Quantifying Mineral Dust Mass Budgets: Terminology, Constraints, and Current Estimates, Eos Trans. AGU, 85(48), 509–512, doi:10.1029/2004EO480002, 2004

[5] Solomos S., Kallos G., Kushta J., Astitha M., Tremback C., Nenes A., Levin Z. An integrated modeling study on the effects of mineral dust and sea salt particles on clouds and precipitation, *Atmospheric Chemistry and Physics* 11 (2011) 873-892

[6] Kallos, G., Astitha, M., Katsafados, P., and Spyrou, C.: Long-Range Transport of Anthropogenically and Naturally Produced particulate matter in the Mediterranean and North Atlantic: Current State of Knowledge, *J. Appl. Meteorol. Clim.*, 46, 1230–1251, doi:10.1175/JAM2530.1, 2007.

[7] Seinfeld JH, Pandis SN (1998) Atmospheric Chemistry and Physics From Air Pollution to Climate Change, John Wiley & Sons, Inc, New York, 363-381

[8] Perrone MR, Turnone A, Buccolieri A, Buccolieri G Particulate matter characterization at a coastal site in south-eastern Italy. *Journal of Environ Monitoring* 8 (2005) 183-190.

[9] Avila A, Alarcon M Relationship between precipitation chemistry and meteorological situations at a rural site in NE Spain. *Atmospheric Environment* 33 (1999) 1663–1677.

[10] Glavas S, Moschonas N Origin of observed acidic–alkaline rains in a wet-only precipitation study in a Mediterranean coastal site, Patras, Greece. *Atmospheric Environment* 36 (2002) 3089–3099

[11] Yoboue V, Galy-Lacaux C, Lacaux JP, Silue S Rainwater Chemistry and Wet Deposition over the Wet Savanna Ecosystem of Lamto (Côte d'Ivoire). *Journal of Atmospheric Chemistry* 52 (2005)117-141

[12] Luria M, Peleg M, Sharf G, Tov-Alper DS, Spitz N, Ben Ami Y, Gawii Z, Lifschitz B, Yitzcaki A, Seter I (1996) Atmospheric sulphur over east Mediterranean region. *Journal of Geophysical Research* 101:25917-25930

[13] Tost, H., Jöckel, P., Kerkweg, A., Pozzer, A., Sander, R., and Lelieveld J., Global cloud and precipitation chemistry and wet deposition: tropospheric model simulations with ECHAM5/MESSy1. *Atmospheric Chemistry and Physics* (7) (2007) 2733-2757

[14] Stevenson, D.S., Dentener, F.J., Schulz, M.G., Ellingsen, K., Van Noije T.P.C., Wild, O., Zeng, G., Amann, M., Atherton, C.S., Bell, N., Bergmann, D.J., Bey, I., Butler, T., Cofala, J., Collins, W.J., Derwent, R.G., Doherty, R.M., Drevet, J., Eskes, H.J., Fiore, A.M., Gauss, M., Hauglustaine, D.A., Horowicz, L.W., Isaksen, I.S.A., Krol, M.C., Lamarque, J.-F., Lawrence, M.G., Montanaro, V., Mu¨ ller, J.-F., Pitari, G., Prather, M.J., Pyle, J.A., Rast, S., Rodriguez, J.M., Sanderson, M.G., Savage, N.H., Shindell, D.T., Strahan, S.E., Sudo, K., Szopa, S., 2006. Multimodel ensemble simulations of present-day and near-future tropospheric ozone. *Journal of Geophysical Research 111*, D08301. doi:10.1029/2005JD006338.

[15] Fowler D., Pilegaard K., Sutton M.A., Ambus P., Raivonen M., Duyzer J., Simpson D., Fagerli H., Fuzzi S., Schjoerring J.K., Granier C., Neftel A., Isaksen I.S.A., Laj P., Maione M., Monks P.S., Burkhardt J., Daemmgen U., Neirynck J., Personne E., Wichink-Kruit R., Butterbach-Bahl K., Flechard C., Tuovinen J.P., Coyle M., Gerosa G., Loubet B., Altimir N., Gruenhage L., Ammannl C., Cieslik S., Paoletti E., Mikkelsen T.N., Ro-Poulsen H., Cellier P., Cape J.N., Horváth L., Loreto F., Niinemets Ü., Palmer P.I., Rinne J., Misztal P., Nemitz E., Nilsson D., Pryor S., Gallagher M.W., Vesala T., Skiba U., Brüggemann N., Zechmeister-Boltenstern S., Williams J., O'Dowd C., Facchini M.C., De Leeuw G., Flossman A., Chaumerliac N., Erisman J.W., Atmospheric composition change: Ecosystems–Atmosphere interactions, Review, *Atmospheric Environment* 43 (2009) 5193–5267

[16] EMEP, 2000. Extracted from http://www.emep.int/

[17] Avila A., Rodá F., Assessing decadal changes in rainwater alkalinity at a rural Mediterranean site in the Montseny Mountains (NE Spain), *Atmospheric Environment* 36 (2002) 2881–2890

[18] Kubilay N, Ničković S, Moulin C, Dulac F (2000) An illustration of the transport and deposition of mineral dust onto the eastern Mediterranean. *Atmospheric Environment* 34:1293-1303

[19] Özso E, Kubilay N, Ničković S, Moulin C (2001) A hemispheric dust storm affecting the Atlantic and Mediterranean in April 1994: Analyses, modeling, ground-based measurements and satellite observation. *Journal of Geophysical Research* 106, 18439-18460.

[20] Vukmirović, Z., Unkašević, M., Lazić, L., Tošić, I., Rajšić, S., Tasić, M., 2004. Analysis of the Saharan dust regional transport. Meteorology and Atmospheric Physics. *Meteorology and Atmospheric Physics* 85 (4) 265-273

[21] Rodríguez S, Querol X, Alastuey A, Viana M-Mar, Mantilla E (2003) Events affecting levels and seasonal evolution of airborne particulate matter concentrations in the Western Mediterranean. *Environmental Science and Technology* 37:216-222

[22] Draxler RR, Rolph GD (2003) HYSPLIT (HYbrid Single-Particle Lagrangian Integrated Trajectory) Model access via NOAA ARL READY Website (http://www.arl.noaa.gov/ready/hysplit4.html). NOAA Air Resources Laboratory, Silver Spring, MD

[23] Draxler RR, Hess GD (1997) *Description of the Hysplit_4 modeling system,* NOAA Tech Memo ERL ARL-224, Dec, 24pp

[24] Draxler RR, Hess GD (1998) An Overview of the Hysplit_4 Modeling System for Trajectories, Dispersion, and Deposition, *Australian Meteorological Magazine* 47:295-308

[25] Hsu Y, Holsen TM, Hopke PK (2003) Locating and Quantifying PCB Sources in Chicago: Receptor Modeling and Field Sampling. *Environmental Science and Technology* 37 (4): 681-690

[26] Mihajlidi-Zelić A, Deršek-Timotić I, Relić D, Popović A, Đorđević D (2006) Contribution of marine and continental aerosols to the content of major ions in the precipitation of the central Mediterranean. *Science of the Total Environment* 370 (2-3):441-451

[27] Đorđević D., Tošić I., Unkašević M., Đurašković P., Water-soluble main ions in precipitation over the southeastern Adriatic region: chemical composition and long-range transport, *Environmental Science and Pollution Research* 17 (9) (2010) 1591-1598

[28] Djordjevic D., Radmanovic D., Mihajlidi-Zelic A., Vukovic T (2002a) Intercomparisons of Na, K, Mg and Ca concentrations in the precipitation from the 18th intercomparison of analytical methods within EMEP (2000), 5th International conference of Balkan Environmental Association (B.EN.A) on Transboundary Pollution, 7 – 10 November, Belgrade.

[29] Đorđević D, Radmanović D, Mihajlidi-Zelić A, Vuković T (2002b) Intercomparasion of Na, K, Mg and Ca in the precipitation from the 23rd WMO-GAV laboratory intercomparasion (2000). *Journal of Environmental Protection and Ecology* (3) 4: 823 - 827

[30] Becagli S., Proposito M., Benassai S., Gragnani R., Magand O., Traversi R., Udisti R., Spatial distribution of biogenic sulphur compounds (MSA, $nssSO_4^{2-}$) in the northern Victoria Land–Dome C–Wilkes Land area, East Antarctic, *Annals of Glaciology* 41 2005.

In: Precipitation: Prediction, Formation, Environmental Impact ISBN: 978-1-62100-447-9
Editors: Henry Dohring, Jeremy Dixon

Chapter 6

COPING WITH DAILY PRECIPITATION ERRORS OF REGIONAL CLIMATE MODELS

Matthias Jakob Themeßl*, Martin Suklitsch and Andreas Gobiet
Wegener Center for Climate and Global Change and Institute for Geophysics, Astrophysics, and Meteorology, University of Graz, Austria

1. INTRODUCTION

Precipitation amount is one of the most demanded meteorological parameters for climate change impact studies (e.g., Swart and Pagé, 2010). It represents an important input for studies across various economic sectors such as agriculture, hydrology, energy and tourism. Furthermore, precipitation acts as a trigger for extreme events such as landslides and floods (Frei et al., 2000; Jomelli et al., 2009). However, the correct simulation of precipitation intensity and the sequence of wet and dry conditions at the daily and sub-daily scale is very challenging. Homogenous Markov chains represent one of the first attempts to capture and reproduce daily precipitation characteristics and are still common practice in statistical modeling (von Storch and Navarra, 1991; Wilby and Wigley, 1997). Besides, climate models nowadays are the state-of-the-art tools for dynamically simulating historic as well as future climate conditions in a physically consistent way. Climate models are mathematical representations of the climate system with its subsystems, their processes and interactions (Baede et al., 2001). Their development, enhancement and availability went along with advances in high performance computing over the last decades. Global climate models (GCMs), which feature horizontal resolutions of about 200 km, appropriately simulate global distributions of, e.g., temperature, precipitation, radiation, wind, ocean temperature, currents and ice cover as well as important patterns of climate variability such as the Hadley cell, monsoon systems and storm tracks (Randall et al., 2007). However, for regional or even finer scale climate assessments GCM output is too coarse. Furthermore, they neglect or parameterize sub-grid-scale forcings and processes. This leads to biases in the simulation of regional climate conditions (Giorgi et al., 2001) and, e.g., unreliable temporal output at the

* Corresponding author: matthias.themessl@uni-graz.at.

daily scale (Prudhomme et al., 2002). Such spatially and temporally fine resolved data is though essential for the climate impact community. In order to bridge the gap towards useful daily data at the regional scale, GCM information is cascaded down (downscaled) either dynamically or statistically (e.g., Maraun et al., 2010).

Empirical-statistical downscaling (ESD) relates large scale, e.g., synoptic conditions, taken from GCMs or reanalysis data, to observations via a statistical model (e.g., Benestad et al., 2008). By this means, climate information can be generated at the point-scale. Furthermore, if the statistical model is calibrated on climate model data, ESD additionally functions as a climate model error correction (compare e.g., Maraun et al., 2010). Dynamical downscaling, in contrast, nests a regional climate model (RCM) into the GCM over a limited area with higher spatial resolution (10 km to 50 km grid spacing). The RCM, consequently, benefits from its resolution by a more accurate representation of regional climate forcings (e.g., orography, land use, or land-sea contrasts) and the ability to resolve processes, which are parameterized in GCMs (Giorgi and Mearns, 1991, 1999; Laprise, 2008; Maraun et al., 2010 and references therein). Hence, RCMs have proven to add regional information compared to GCM output (e.g., Wang et al., 2004; Feser et al., 2006; Laprise, 2008) including more plausible mesoscale circulation patterns (Buonomo et al., 2007) or estimations of extremes (e.g., Schmidli et al., 2007; Christensen and Christensen, 2007).

By combining both downscaling pathways, this study presents a methodological approach, which aims at providing as accurate as possible fine scale climate information for the climate impact community. In the following, Section 2 describes state-of-the-art RCM simulations over Europe and state-of-the-art observational data for comparison, Section 3 evaluates the RCM simulations and indicates their error characteristics for daily precipitation amount, Section 4 introduces and evaluates empirically based statistical post-processing techniques to mitigate potential errors, before Section 5 summarizes the key findings and draws conclusions.

2. DATA

Within the last years the availability of RCM data for Europe rapidly increased thanks to projects such as PRUDENCE (Christensen and Christensen, 2007) and ENSEMBLES (van der Linden and Mitchell, 2009). The dataset will be further complemented in the currently ongoing CORDEX project (http://wcrp.ipsl.jussieu.fr/SF_RCD_CORDEX.html). ENSEMBLES contained a coordinated endeavor to produce regional climate scenarios for Europe and West Africa. In the course of the project several improvements were made regarding regional climate downscaling. For example, the horizontal resolution of the simulations was increased from 50 km (PRUDENCE) to 25 km. Furthermore, transient climate simulations were produced, i.e., continuous simulations from the 1950s to the year 2100 for a well filled RCM/GCM matrix, which were rare at the time the project started.

In this study, 16 regional climate simulations, driven with ERA-40 reanalysis data (Uppala et al., 2000), from the ENSEMBLES project will be evaluated. The RCMs that are used for the simulations are listed in Table 1.

Table 1. List of evaluated ENSEMBLES RCMs and their operators

RCM Acronym	Simulation performed by
RCA3	Community Climate Change Consortium for Ireland (C4I)
ALADIN	Meteo France (CNRM)
HIRHAM	Danish Meteorological Institute (DMI)
CLM	Swiss Institute of Technology (ETHZ)
CLM	Forschungszentrum Geesthacht GmbH (GKSS)
HadRM3Q0	UK Met Office, Hadley Centre for Climate Prediction and Research (HC)
HadRM3Q3	UK Met Office, Hadley Centre for Climate Prediction and Research (HC)
HadRM3Q16	UK Met Office, Hadley Centre for Climate Prediction and Research (HC)
RegCM	The Abdus Salam Intl. Centre for Theoretical Physics (ITCP)
RACMO	The Royal Netherlands Meteorological Institute (KNMI)
HIRHAM	Norwegian Meteorological Institute (METNO)
REMO	Max Planck Institute for Meteorology (MPI)
CRCM	Consortium on regional climatology and adaptation to climate change (OURANOS)
RCA	Swedish Meteorological and Hydrological Institute (SMHI)
PROMES	Universidad de Castilla-La Mancha (UCLM)
GEMLAM	Environment Canada (EC)

In addition to the model simulations, the ENSEMBLES project provided a multi-decadal gridded observation dataset called E-OBS (Haylock et al., 2008) at a horizontal resolution of roughly 25 km. It is based on the most complete collection of station data over wider Europe (Klok and Klein-Tank, 2009). E-OBS specifically aims at the evaluation of regional climate models as the gridded data represent areal averages, equal to RCM outputs, rather than point-scale data (e.g., Déqué, 2007). E-OBS contains near surface (2 m) air temperature (daily mean, minimum and maximum) and precipitation amount (daily sums) as well as uncertainty estimates for all parameters. In the course of time, E-OBS was updated including more stations and a longer time range. In its most recent version 4 (v4), E-OBS covers the period from 1958 to 2010 and also contains daily fields of mean sea level pressure (van den Besselaar et al., 2011).

3. Error Characteristics of Daily Precipitation Amount from State of the Art RCMs Over Europe

All considered regional climate simulations are evaluated against E-OBS v4 on daily basis between 1961 and 2000 and then combined to a multi-model ensemble (MME). The MME bias (presented as the median bias of the MME) is given as an estimate for the errors of current state-of-the-art RCMs. Besides, an inter-model error range of the MME is defined as the interval between the 5th and the 95th percentile in order to retain as much data as possible in the ensemble while disregarding outliers. Furthermore, the MME is evaluated for wet-day frequency (number of days with a precipitation sum ≥ 1 mm/day) and precipitation intensity (mean precipitation amount on wet days). These two parameters are considered in this study in order to take into account the at zero-bounded, non-symmetric distribution of daily precipitation.

Interpreting the results it should be kept in mind that the evaluation methodology regards the observational dataset as an error-free reference. According to Hofstra et al. (2009) E-OBS is of higher quality for mean values and temperature parameters than for extremes and precipitation amount.

In Figure 1 the relative MME bias is illustrated on seasonal basis. This comparison reveals a wet bias on average for Europe throughout the entire year. The mean bias is largest in spring (MAM, Figure 1b) with +41.6 % and ranges between about —20 % about +110 % on smaller scales (compare the 5th and 95th percentiles of the error maps). For single models the small scale biases can also reach up to +250 % (not shown). The MME bias shows a distinct annual cycle as well as distinct regional variability. The area average bias increases from winter to spring and then decreases until autumn. The decrease can be related to a widespread dry area up to —35 % precipitation underestimation in south-eastern Europe in summer and in the Mediterranean region in autumn. This summer dry bias pattern resembles the respective error pattern described in Hagemann et al. (2004) and Jacob et al. (2007). However, interestingly the rest of the Mediterranean region in summer indicates a distinct wet-bias $\geq$ 35 %.

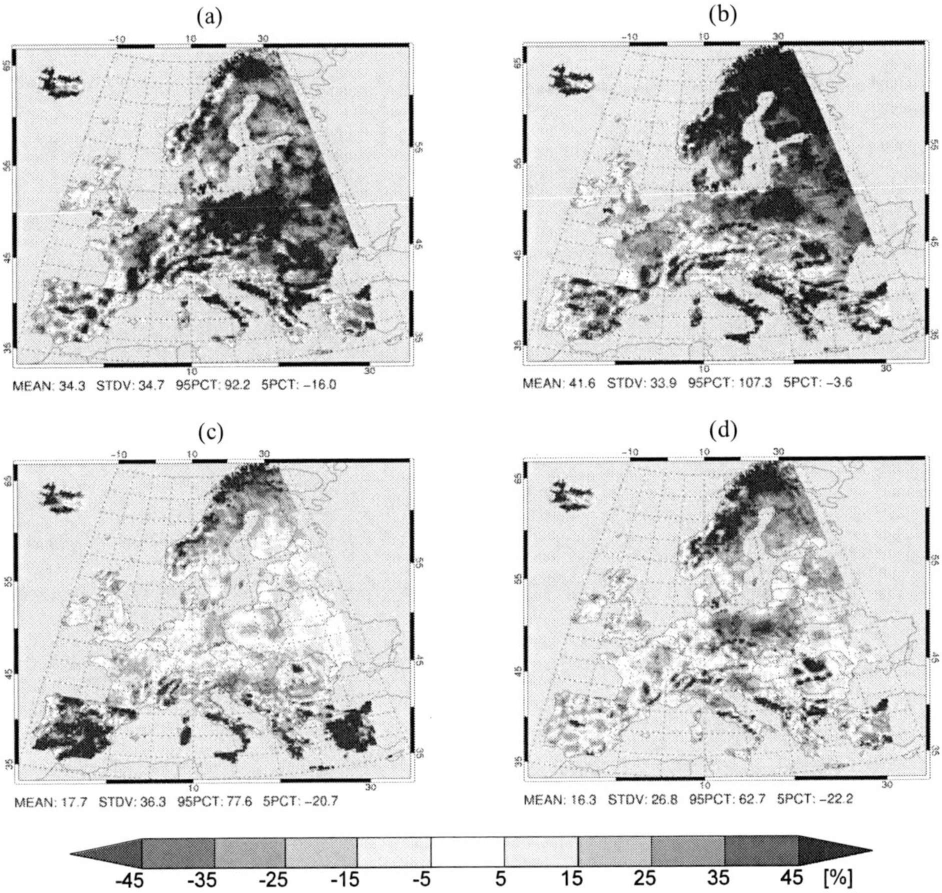

Figure 1. Median of the relative seasonal precipitation bias of the ENSEMBLES multi-model ensemble (MME). (a) winter (DJF), (b) spring (MAM), (c) summer (JJA), (d) autumn (SON). STDV: standard deviation; 95PCT: 95th percentile; 5PCT 5th percentile.

Independent of the season, mountain ranges are visible in the MME bias pattern, predominantly as enhanced wet biases. This obviously orographically induced precipitation error is rather related to model difficulties with orographic lifting than to increases in the horizontal flow (Christensen et al., 1997, Suklitsch et al., 2010). Referring to Hofstra et al. (2009), these errors may also be related to inaccurate observation data.

Concerning the accordance in the MME, seasonal inter-model MME error ranges in Table 2 indicate that the MME in general ranges between about −31 % and +95 % depending on the season. However, the MME particularly accords concerning the spring wet-bias indicated by a strongly reduced 5th percentile at −6.8 %.

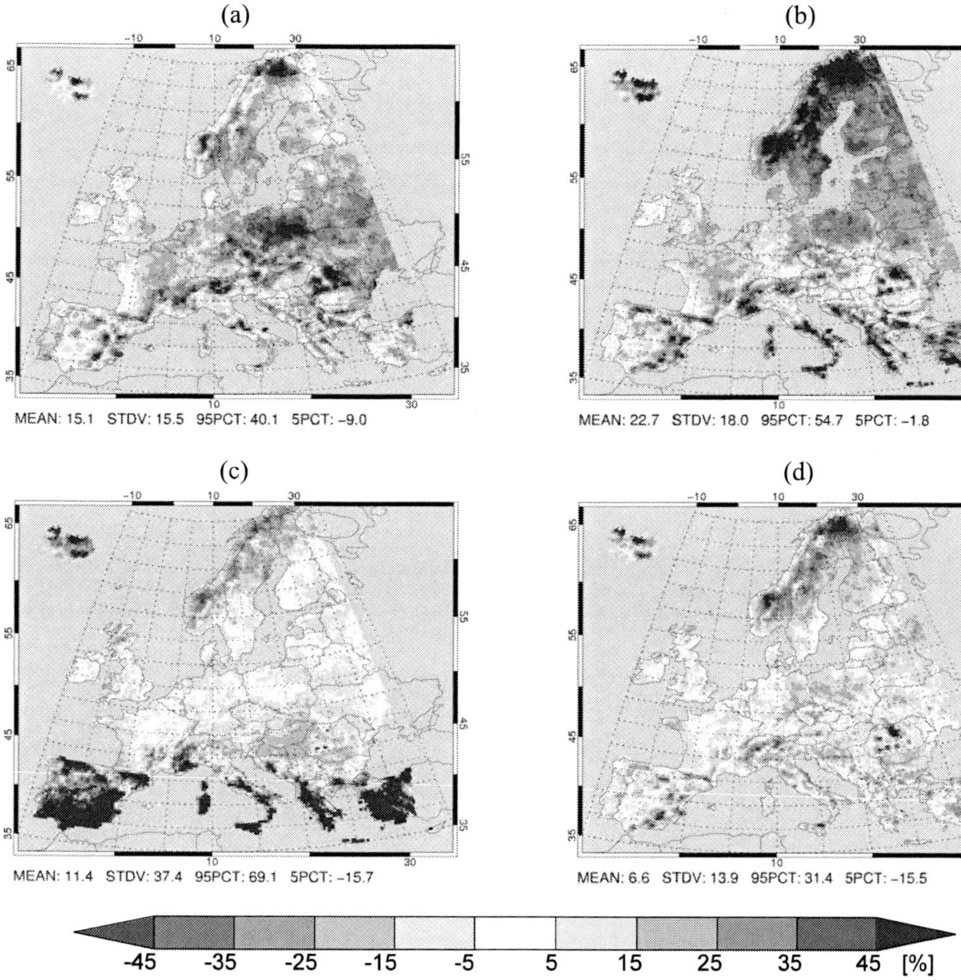

Figure 2. Median of the relative seasonal bias of precipitation frequency of the ENSEMBLES multi-model ensemble (MME). (a) winter (DJF), (b) spring (MAM), (c) summer (JJA), (d) autumn (SON). Abbreviations in the panels' footers are defined in Figure 1.

The MME precipitation frequency bias, depicted in Figure 2, shows comparable error characteristics in the annual cycle as well as in the spatial pattern as precipitation amount. The maximum area average frequency bias amounts to +22.7% in spring (MAM) and varies between about −15 % and about +70 % for smaller regions. Outstanding is the homogenous wet bias in the Mediterranean region in summer. The corresponding MME inter-model seasonal error ranges in Table 2 underline the strong Mediterranean wet bias in summer.

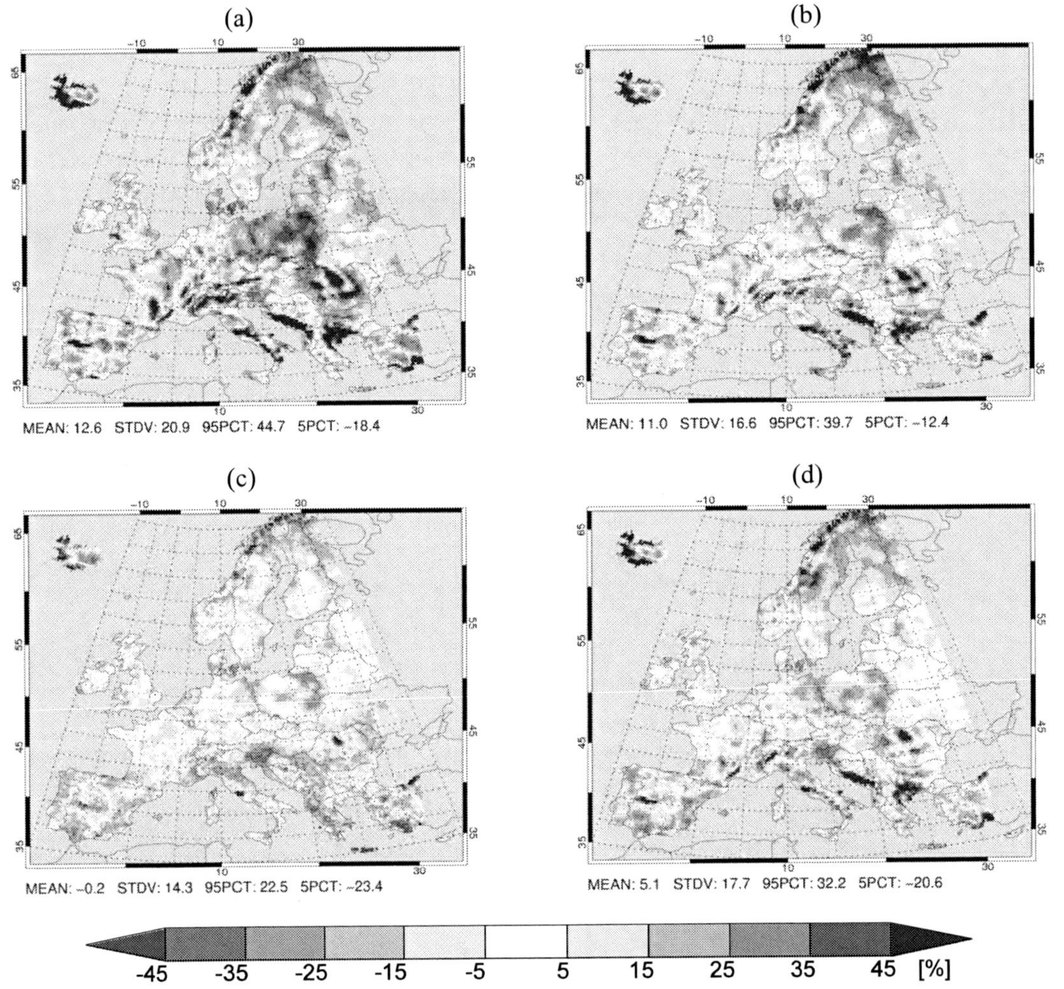

Figure 3. Median of the relative seasonal bias of precipitation intensity of the ENSEMBLES multi-model ensemble (MME). (a) winter (DJF), (b) spring (MAM), (c) summer (JJA), (d) autumn (SON). Abbreviations in the panels' footers are defined in Figure 1.

In Figure 3 we show the seasonal relative MME bias for precipitation intensity. The bias features more spatial and temporal variability than precipitation amount and precipitation frequency and is slightly negative in summer on average (Figure 3c). Particularly in winter, but also in spring and autumn biases are related to the orography. The rather homogenous intensity underestimation over mid and southern Europe in summer indicates that the RCMs might have deficiencies to account for convective precipitation events. However, the respective MME error ranges, given in Table 2, do not accord in the presented seasonality of the precipitation intensity bias. The respective ranges, which are firstly smaller than for precipitation sums and precipitation frequency, remain rather constant over the year.

Table 2. Areal averages over Europe of the inter-model error range of the MME bias for the three evaluated parameters (columns) separated by season (rows). Units: [%]

	Precipitation amount			Precipitation frequency			Precipitation intensity		
Season	MME mean	5th percentile	95th percentile	MME mean	5th percentile	95th percentile	MME mean	5th percentile	95th percentile
Winter (DJF)	34.3	−15.4	86.1	15.1	−19.8	43.5	12.6	−13.0	38.6
Spring (MMA)	41.6	−6.8	95.2	22.7	−12.7	57.6	11.0	−14.5	37.1
Summer (JJA)	17.7	−30.8	90.6	11.4	−28.0	92.0	−0.2	−26.2	29.8
Autumn (SON)	16.3	−30.6	57.7	6.6	−27.0	44.6	5.1	−24.8	28.0

4. Ways to Mitigate RCM Error Characteristics

From the previous section it becomes clear that state-of-the-art RCMs are significantly biased compared to observations. These errors complicate the direct application of RCM results to climate impact assessment or decision making since they lead to biased estimates of climate change impacts. Wood et al. (2004) stated that the minimum standard for any useful downscaling procedure is that ‘the historic (observed) conditions must be reproducible’. For the purpose of providing error-corrected climate data, empirical-statistical post-processing has been proposed, e.g., by Widmann et al. (2003) for large scale reanalysis data and Wood et al. (2004) for GCMs, and recently been evaluated by, e.g., Déqué (2007), Graham et al. (2007), Leander and Buishand (2007), Dobler and Ahrens (2008), Piani et al. (2009) and Themeßl et al. (2011a) for daily precipitation amount simulated by RCMs. Based on the concept of model output statistics (MOS; Maraun et al., 2010), such statistical post-processing methods calibrate a correction function between the climate model output and observations, which forces the RCM simulation towards observations, thus corrects it, assuming that the observations represent the truth. If the underlying observations feature a higher spatial resolution, they implicitly downscale the simulations in addition. Therefore, they are also called “downscaling and error correction methods” (DECMs). Contrary to the traditional empirical-statistical downscaling approach of perfect prog(nosis) (e.g., Maraun et al., 2010), where the statistical model is fitted between observations at different scales and thus is applicable to any climate model, the calibrated correction and downscaling function in MOS is only valid for the respective climate model.

Themeßl et al. (2011a) evaluated seven different empirical-statistical DECMs for daily precipitation amount as well as for derived parameters as precipitation frequency, intensity and moderate extremes. Their evaluation was based on a 10 km resolved RCM hindcast over the orographically demanding European Alpine region. They concluded that Quantile Mapping (QM) systematically outperformed all other DECMs although Local Intensity Scaling and analog methods resulted in nearly similar error correction potential. Based on their findings, as well as studies of, e.g., Boé et al. (2007), Déqué (2007) and Dobler and Ahrens (2008) QM is proposed also in this study as the DECM of choice for illustrative purposes. QM can be defined as a probability-based correction procedure (compare Figure 4 right panel), which allows for correction of errors in the entire precipitation distribution. If QM is based on empirical cumulative density functions (ecdf), thus applied in a non-parametric way, it is in general applicable to any meteorological parameter without a priori knowledge of the underlying probability distribution. Furthermore, in the context of daily precipitation this application is feasible without the separation between dry and wet-days.

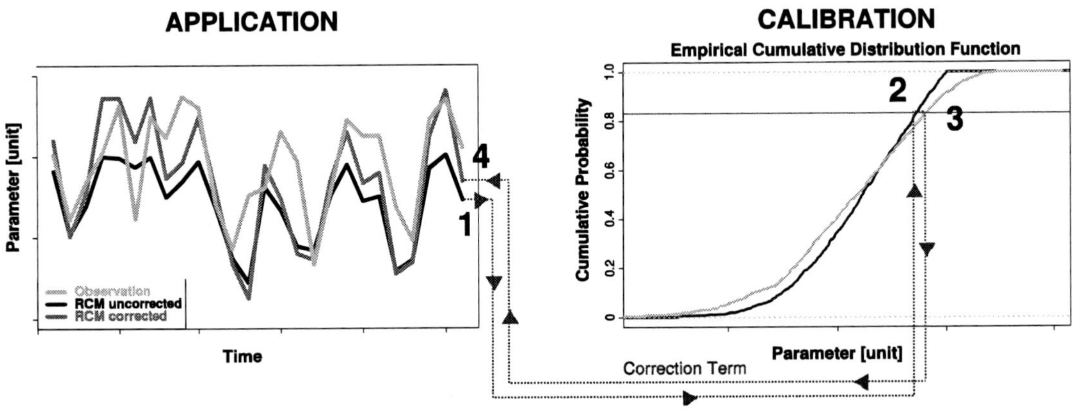

Figure 4. Scheme of QM. An uncorrected daily value from a climate model (1) is related to its respective empirical cumulative distribution function (ecdf; Wilks, 2006) for the same day of the year in the calibration period (2). The difference between the absolute value associated with the corresponding observational probability (3) and the uncorrected climate model output from (1) is taken as the correction value and added to the uncorrected value to obtain the corrected one (4).

In this study, QM is applied on daily basis, using a 31-day moving window for ecdf calibration. This way, the seasonality of the error characteristics can be taken into account separately for each grid cell, while obtaining a sufficient large sample for calibration. Furthermore, this implementation allows correcting spatially varying error characteristics. However, spatial covariances are not taken into account. In addition, QM incorporates a frequency adaptation (Themeßl et al., 2011b), which allows to accurately correct conditions when there are less dry days in the observations than in the model results. This situation, however, is uncommon in RCMs as they tend to overestimate precipitation frequency (compare Figure 2), which is known as “drizzling problem” (e.g., Gutowsky et al., 2003). Finally, QM here constantly extrapolates the correction value of the historic minimum or maximum onto values outside the calibration range. This procedure was shown to produce “new extremes” without deterioration and mostly with improvement of the original RCM quality (Themeßl et al., 2011b).

Figure 5 compares one selected ENSEMBLES RCM — CLM (Böhm et al., 2006), applied by the Swiss Federal Institute of Technology, Zurich (ETHZ) — in its original, uncorrected version to the respective corrected data regarding annual precipitation amount (panels a and b) as well as annual precipitation frequency (panels c and d) and intensity (panels e and f). Similar to Section 3 the evaluation is based on daily data for the period 1961 to 2000 and performed against E-OBS v4. Since the observational and simulated data feature nearly the same horizontal resolution, no further downscaling is achieved. Thus, the evaluation focuses on the correction potential of spatial and temporal errors for the parameters introduced in Section 3 as well as on the correction potential for intensity-dependent error characteristics regarding the entire precipitation distribution. The corrected data are obtained using a decadal cross validation approach (Themeßl et al., 2011a).

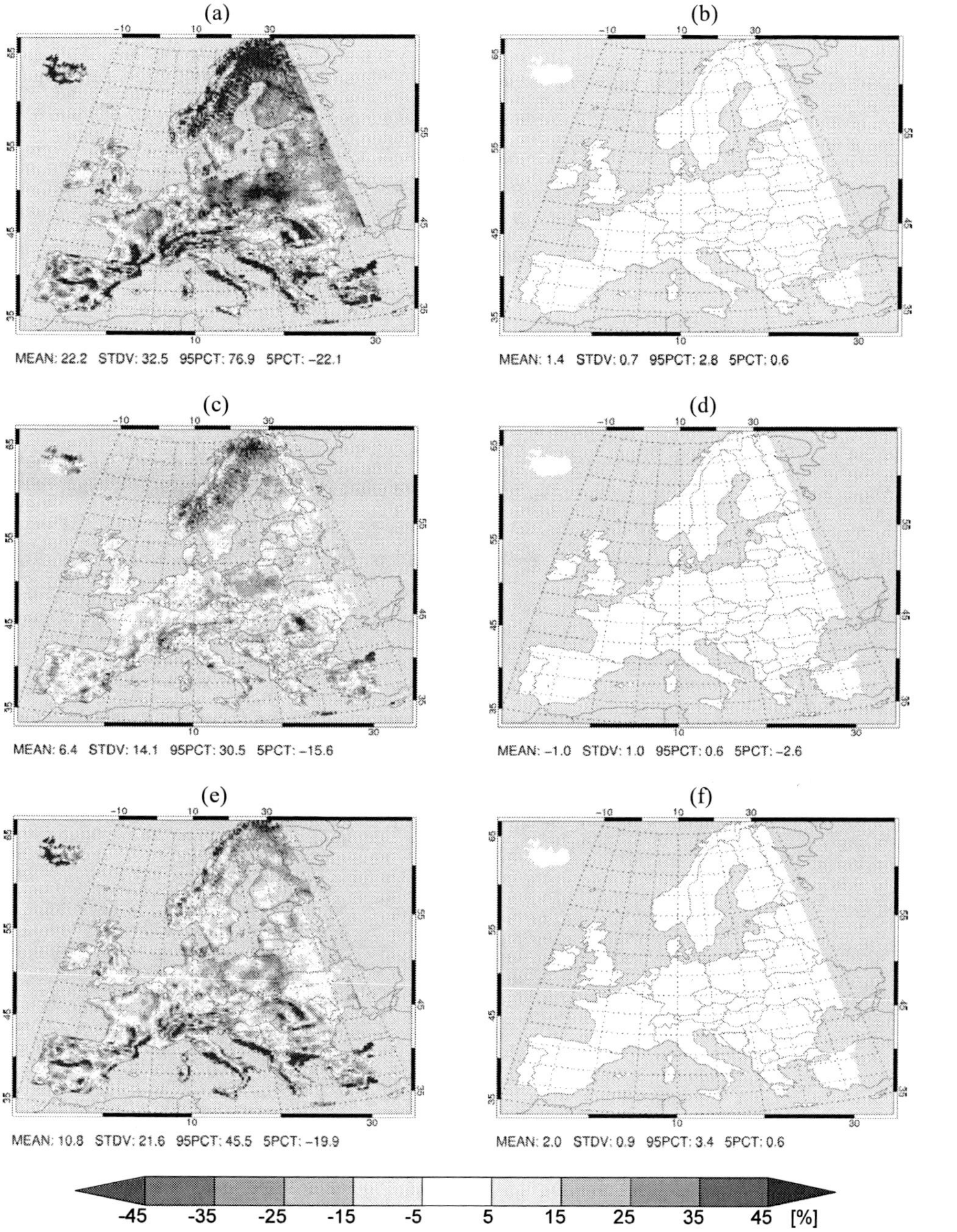

Figure 5. Annual relative bias of uncorrected precipitation amount (a) corrected precipitation amount (b), uncorrected precipitation frequency (c), corrected precipitation frequency (d), uncorrected precipitation intensity (e) and corrected precipitation intensity (f) from the CLM hindcast between 1961 and 2000 compared to E-OBS v4. Abbreviations in the panels' footers are defined in Figure 1.

Annual uncorrected precipitation in Figure 5a indicates that CLM is too wet, except over the Iberian Peninsula, the British Isles, parts of central to southeastern Europe, and over Turkey. The area average bias amounts to about +20. % and reaches up to about +75 % for smaller regions. With error correction the regionally highly variable bias is drastically reduced to +1.4 % on average and to about +3 % on smaller scales. Regarding Figure 5c, CLM simulates too many wet days, in average +6.4 % more than the observational reference shows. The bias features a north-south gradient from overestimation in the North to underestimation in the South, with exceptions in some mountain ranges. After error correction, the bias is strongly reduced similar to mean precipitation by at least one order of magnitude of the uncorrected bias. Also concerning precipitation intensity this error correction potential is visible in Figure 5e and 5f.

Box and Whisker plots in Figure 6 focus on the temporal characteristics of the area average bias. On the seasonal scale, the uncorrected CLM in Figure 6 (left boxes and whisker in each season) varies in accordance with the description given along with Figure 1, although in this case the minimum median bias is found in summer. The error spread for this RCM, illustrated by the 5th and 95th percentile in the Box and Whisker plots, is similar for winter, spring and autumn but amplified in summer. This amplification indicates higher variability in this season, likely due to already mentioned problems with convection in the RCM.

QM reduces the temporal bias to near zero independent of the season (right boxes and whiskers in each season in Figure 6). However, contrary to spatial bias maps in Figures 1-3 and Figure 5, the range of the temporal differences remains unchanged. This is related to the fact that QM does not alter the temporal characteristics of the time series, i.e. a very wet day still remains wet after correction.

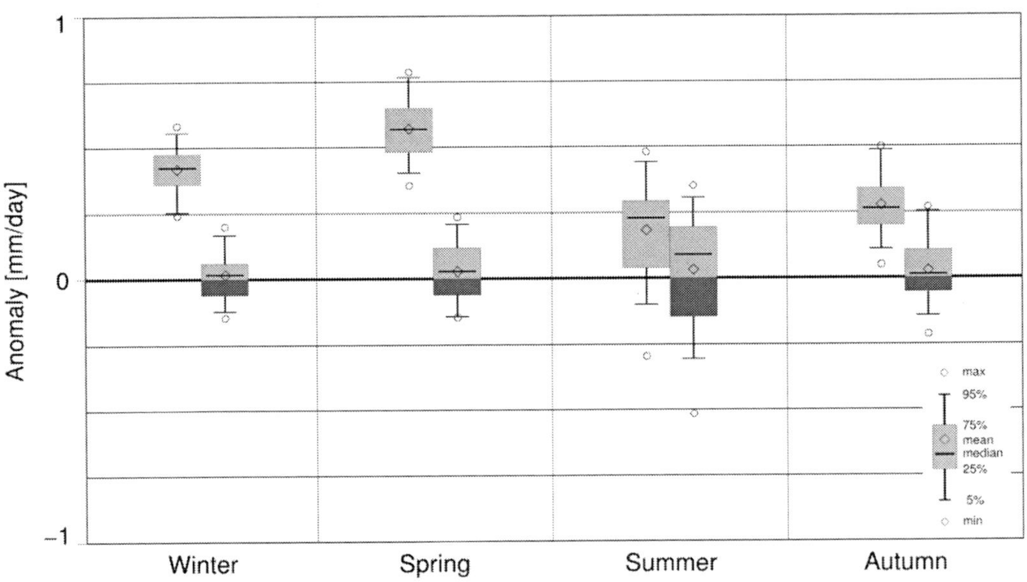

Figure 6. Box and Whisker-Plots of the seasonal bias of uncorrected (left in each season) and corrected (right in each season) daily precipitation amount for entire Europe between 1961 and 2000 compared to E-OBS v4.

Concluding the results section, we compare selected seasonal and regional distributions of the uncorrected and corrected data to observations in Figure 7. The distributions are limited to 3 mm/day for visualization purposes. The shown European sub-regions are defined in Christensen and Christensen (2007). Both cases nicely illustrate the already mentioned drizzling problem of RCMs with too few dry days and too many days with small precipitation sums. Although the observed precipitation characteristics as well as the error characteristics are completely different for winter precipitation in Scandinavia (Figure 7a) and for summer precipitation in the Mediterranean region (Figure 7b), the corrected data accord in the frequency of dry days as well as the precipitation intensity distribution with the observation. Thus, the corrected data also feature the same variability as the observations.

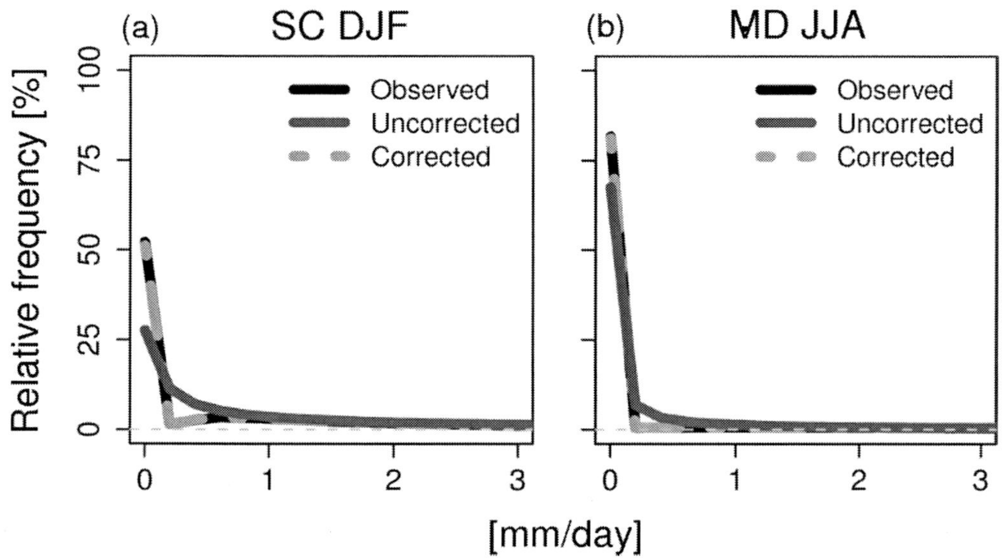

Figure 7. Selected seasonal distributions of daily precipitation sums between zero precipitation and 3 mm/day for Scandinavia (SC) (a), and for the Mediterranean region (MD) (b) between 1961 and 2000. DJF: winter season; JJA: summer season.

CONCLUSION

Accurate time series of daily precipitation are an essential input for various climate impact assessments. Regional climate models (RCMs) represent the state-of-the-art tools for providing physically consistent regional climate conditions over limited areas at the daily scale. Within the last decades their availability and accuracy has increased. Based on the latest ensemble of RCM simulations for daily precipitation over Europe, taken from the ENSEMBLES project, we show however that the requirements of impact researchers for useful climate data that accurately reproduce the historic conditions are not yet met. The evaluation of a multi model ensemble (MME) of 16 ERA-40 driven RCM simulations for the period 1961 to 2000 reveals significant errors in areal average precipitation sum, precipitation frequency as well as precipitation intensity. In general, the MME is too wet. Depending on the season the European spatial mean of the MME varies between about +15 % and +40 % for precipitation sum, between about +5 % and +20 % for precipitation frequency and between about 0 % and +10 % precipitation intensity. The inter-model error range of the MME precipitation bias, defined as the range between the 5th and 95th percentile in the MME, yields a spread between about —30 % underestimation and about +95 % overestimation, with variations up to 40 % on seasonal basis. The respective MME error ranges for precipitation frequency and especially for intensity are smaller compared to mean precipitation and concerning precipitation intensity indicate smaller inter-annual variability.

The error pattern for precipitation amount features a distinct spatial characteristic. Whereas in winter and spring the MME results in a pronounced precipitation overestimation for large parts of Europe, in summer and autumn widespread regions in south-eastern Europe and the Mediterranean are too dry. Furthermore, the error pattern clearly indicates an elevation dependency as nearly all mountainous regions show distinguishable positive biases compared to flatter regions. In smaller regions (neglecting single outlier grid cells), the biases range from about —20 % to +110 for the MME and can amount up to about +250 % for single models on the seasonal scale. This spatial variability demonstrates the still existing deficiencies of RCMs concerning the spatially correct simulation of precipitation events, particularly for demanding regions as the Alps. Precipitation frequency and intensity feature partly comparable error patterns as mean precipitation amount. Nevertheless, the respective precipitation frequency error characteristics are less pronounced despite an outstanding wet-bias in the Mediterranean region in summer and the precipitation intensity error pattern shows more regional variability.

Assessing the presented RCM's skill it should however be considered that the evaluation methodology assumes the observational dataset to be error-free which here at least for mountainous regions remains questionable.

In order to provide more robust climate information for climate impact research as well as decision making, the post-processing of RCM results using empirical-statistical techniques is advisable. We demonstrated the potential of combining one selected RCM (CLM) with an empirical-statistical downscaling and error correction method (DECM), namely Quantile Mapping (QM). Based on a decadal cross validation approach between 1961 and 2000 QM results in strongly improved climatologies. Independent of regionally and seasonally varying errors, corrected precipitation biases are at least reduced by one order of magnitude, with remaining errors mostly near zero. Furthermore, the entire distribution, thus also precipitation frequency and intensity are appropriately corrected. By this means, the data becomes useful, e.g., for calibrating climate impact models. Previous studies confirm QM's applicability to other RCMs, further parameters such as temperature as well as for correcting extremes (Piani et al., 2009, Themeßl et al., 2011a, 2011b). The latter issue is essential for natural hazards estimations since huge daily precipitation amounts can act as triggers for, e.g., landslides.

Besides its depicted advantages it should be noted that QM does not yet improve temporal persistence characteristics as correlation or wet and dry spells. Furthermore, it is applied separately for each grid cell, which in rare cases can lead to spatially incoherent estimations.

DECMs in general can also be used to correct future scenarios assuming that the error correction functions will not change for future periods. For such purposes, the demonstrated error correction potential may however be reduced for single periods due to natural variability. Hagemann et al. (2011) and Themeßl et al. (2011b) discussed the impact of QM on the climate change signal (CCS). Both studies found, that QM can, in some cases even drastically, change the CCS. If these impacts reduce the uncertainty in climate scenarios or even add a new uncertainty component represents scientific work in progress.

ACKNOWLEDGMENTS

The ENSEMBLES data used in this work was funded by the EU FP6 Integrated Project ENSEMBLES (Contract number 505539) whose support is gratefully acknowledged especially the CCLM simulations provided by the ETH Zurich. We furthermore acknowledge the E-OBS dataset from the ENSEMBLES project and the data providers in the ECA&D project (http://eca.knmi.nl) as well as the EU FP6 project CLAVIER where the empirical-statistical procedures have been implemented.

REFERENCES

Baede APM, Ahlonsou E, Ding Y, Schimel D, Bolin B, Pollonais S. The Climate System: An Overview. In: *Climate Change: The Scientific Basis.* Contribution of Working Group I to the Third Assessment Report of the Intergovernmental Panel on Climate Change [Houghton JT, Ding Y, Griggs DJ, Noguer M, van der Linden PJ, Dai X, Maskell K, Johnson CA (eds.)]. Cambridge University Press. Cambridge. UK and New York.

Benestad RE, Hanssen-Bauer I, Chen D. 2008. *Empirical Statistical Downscaling.* World Scientific Publishing Company: New Jersey, London.

Boé J, Terray L, Habets F, Martin E. 2007. Statistical and dynamical downscaling of the Seine basin climate for hydro-meteorological studies. *International Journal of Climatology* 27: 1643—1655.

Böhm U, Kücken M, Ahrens W, Block A, Hauffe D, Keuler, K Rockel B, Will A. 2006. CLM - the climate version of LM: Brief description and long-term applications. *COSMO Newsletter*, 6.

Buonomo E, Jones RG, Huntingford C, Hannaford J. 2007. On the robustness of changes in extreme precipitation over Europe from two high resolution climate change simulations. *Quarterly Journal or the Royal Meteorological Society* 133: 66—81.

Christensen JH, Machenhauer B, Jones RG, Schär C, Ruti PM, Castro M, Visconti G. 1997. Validation of present-day regional climate simulations over Europe:LAM simulations with observed boundary conditions. *Climate Dynamics* 13: 489—506.

Christensen JH, Christensen OB. 2007. A summary of the PRUDENCE model projections of changes in European climate by the end of this century. *Climatic Change* 81:7–30. doi: 10.1007/s10584-006-9210-7

Dobler A, Ahrens B. 2008. Precipitation by a regional climate model and bias correction in Europe and South Asia. *Meteorologische Zeitschrift* 17: 499—509.

Déqué M. 2007. Frequency of precipitation and temperature extremes over France in an anthropogenic scenario: Model results and statistical correction according to observed values. *Global and Planetary Change* 57:16—26

Feser F. 2006. Enhanced Detectability of Added Value in Limited-Area Model Results Separated into Different Spatial Scales. *Monthly Weather Review* 134: 2180—2190.

Frei C, Davies HC, Gurtz J, Schär C. 2000. Climate dynamics and extreme rpecipitation and flood events in Central Europe. *Integrated Assessment* 1: 281—299.

Giorgi F, Mearns LO. 1991. Approaches to the simulation of regional climate change: a review. *Reviews of Geophysics* 29: 191—216.

Giorgi F, Mearns LO. 1999. Introduction to special section: Regional climate modelling revisited. *Journal of Geophysical Research* 104: 6335—6352.

Giorgi F, Hewitson B, Christensen J, Hulme M, Von Storch H, Whetton P, Jones R, Mearns R, Fu C. 2001. Regional Climate Information – Evaluation and Prjection. In: *Climate Change: The Scientific Basis.* Contribution of Working Group I to the Third Assessment Report of the Intergovernmental Panel on Climate Change [Hoghton JT, Ding Y, Griggs DJ, Noguer M, van der Linden PJ, Dai X, Maskell K, Johnson CA (eds.)]. Cambridge University Press. Cambridge. UK and New York.

Graham LP, Andréasson J, Carlsson B. 2007. Assessing climate change impacts on hydrology from an ensemble of regional climate models, model scales and linking methods–a case study on the Lule River basin. *Climatic Change* 81: 293—307.

Gutowski WJ, Decker SG, Donavon RA, Pan Z, Arritt RW, Takle ES. 2003. Temporal-spatial scales of observed and simulated precipitation in central U.S. climate. *Journal of Climate* 16: 3841–3847.

Hagemann S, Machenhauer B, Jones R, Christensen OB, Déqué M, Jacob D, Vidale PL. 2004. Evaluation of Water and Energy Budgets in Regional Climate Models Applied Over Europe. *Climate Dynamics* 23: 547—567.

Hagemann, S, Chen C, Haerter JO, Heinke J, Gerten D, Piani C. 2011. Impact of a statistical bias correction on the projected hydrological changes obtained from three GCMs and two hydrology models *Journal of Hydrometeorology*. 10.1175/2011JHM1336.1. In press.

Haylock, MR., Hofstra N, Klein Tank AMG, Klok EJ, Jones PD, New M. 2008. A European daily high-resolution gridded dataset of surface temperature and precipitation. *Journal of Geophysical Research* 113:1–12. D20119. doi:10.1029/2008JD10201

Hofstra N, Haylock M, New M, Jones PD. 2009. Testing E-OBS European high-resolution gridded data set of daily precipitation and surface temperature. *Journal of Geophysical Research* 114: D21101. doi:10.1029/2009JD011799.

Jacob D, Bärring L, Christensen OB, Christensen JH, de Castro M, Déqué M, Giorgi F, Hagemann S, Hirschi M, Jones R, Kjellström E, Lenderink G, Rockel B, Sánchez E, Schär C, Seneviratne SI, Somot S, van Ulden A, van den Hurk B. 2007. An inter-comparison of regional climate models for Europe: design of the experiments and model performance. *Climatic Change* 81:31—52. doi:10.1007/s10584-006-9213-4.

Jomelli V, Brunstein D, Déqué M, Vrac M, Grancher D. 2009. Impacts of future climatic change (2070–2099) on the potential occurrence of debris flows: a case study in the Massif des Ecrins (French Alps). *Climatic Change* 97 (1-2): 171-191. DOI: 10.1007/s10584-009-9616-0.

Klok EJ, Klein Tank AMG. 2009. Updated and extended European dataset of daily climate observations. *International Journal of Climatolology* 29: 1182–1191. doi:10.1002/joc.1779.

Laprise R. 2008. Regional climate modelling. *Journal of Computational Physics* 227: 3641—3666.

Leander R, Buishand TA (2007) Resampling of regional climate model output for the simulation of extreme river flows. *Journal of Hydrology* 332:487—496.

Maraun D, Ireson A, Wetterhall F, Bachner S, Kendon E, Rust HW, Venema VKC, Widmann M, Chandler RE, Onof CJ, Osborn TJ, Sautner T, Themeßl M, Thiele-Eich I. 2010.

Statistical downscaling and modelling of precipitation. Bridging the gap between dynamical models and the end users. *Reviews of Geophysics* 48,RG3003. doi:10.1029/2009RG000314.

Piani C, Haerter JO, Coppola E. 2009. Statistical bias correction for daily precipitation in regional climate models over Europe. *Theoretical and Applied Climatology*. DOI: 10.1007/s00704-009-0134-9. In press.

Prudhomme C, Reynard N, Crooks S. 2002. Downscaling of global climate models for flood frequency analysis: where are we now. *Hydrological Processes*. 16: 1137—1150. doi: 10.1002/hyp.1054.

Randall DA, Wood RA, Bony S, Colman R, Fichefet T, Fyfe J, Kattsov V,.Pitman A, Shukla J, Srinivasan J, Stouffer RJ, Sumi A, Taylor KE. 2007. Climate Models and Their Evaluation. In: *Climate Change 2007: The Physical Science Basis*. Contribution of Working Group I to the Fourth Assessment Report of the Intergovernmental Panel on Climate Change [Solomon S, Qin D, Manning M, Chen Z, Marquis M, Averyt KB, Tignor M,. Miller HL (eds.)]. Cambridge University Press, Cambridge, United Kingdom and New York, NY, USA

Schmidli J, Goodess CM, Frei C, Haylock MR, Hundecha Y, Ribalaygua J, Schmith T. 2007. Statistical and dynamical downscaling of precipitation: An evaluation and comparison of scenarios for the European Alps. *Journal of Geophysical Research* 112, D04105. DOI: 10.1029/2005JD007026.

Swart R, Pagé C. 2010. *Defining climate modeling user needs: which data are actually required to support impact analysis and adaptation policy development*. American Geophysical Union, Fall Meeting. abstract # GC31E-04.

Suklitsch M, Gobiet A, Truhetz H, Awan NK, Göttel H, Jacob D. 2010. Error Characteristics of High Resolution Regional Climate Models over the Alpine Area. *Climate Dynamics*. doi:10.1007/s00382-010-0848-5.

Themeßl M, Gobiet A, Leuprecht A. 2011a. Empirical-statistical downscaling and error correction of daily precipitation from regional climate models. *International Journal of Climatology*. 31: 1530—1544. doi:10.1002/joc.2168. In press.

Themeßl M, Gobiet A, Heinrich G. 2011b. Empirical-statistical downscaling and error correction of regional climate models and its impacts on the climate change signal. *Climatic Change. Forthcoming.*

Uppala SM, Kållberg PW, Simmons AJ, Andrae U, Da Costa Bechtold V, Fiorino M, Gibson JK, Haseler J, Hernandez A, Kelly GA, Li X, Onogi K, Saarinen S, Sokka N, Allan RP, Andersson E, Arpe K, Balmaseda MA, Beljaars ACM, van De Berg L, Bidlot J, Bormann N, Caires S, Chevallier F, Dethof A, Dragosavac M, Fisher M, Fuentes M, Hagemann S, Hólm E, Hoskins BJ, Isaksen L, Janssen PAEM, Jenne R, Mcnally AP, Mahfouf JF., Morcrette JJ, Rayner NA, Saunders RW, Simon P, Sterl A, Trenberth KE, Untch A, Vasiljevic D, Viterbo P, Woollen J. 2005. The ERA-40 re-analysis. *Quarterly Journal of the Royal Meteorological Society* 131 (612): 2961–3012.

van den Besselaar, EJM, Haylock MR, van der Schrier G, Klein Tank AMG 2011. A European Daily High-resolution Observational Gridded Data set of Sea Level Pressure. *Journal of Geophysical Research*. doi:10.1029/2010JD015468, In press.

van der Linden P, Mitchell JFB (2009) *ENSEMBLES: Climate Change and its Impacts: Summary of research and results from the ENSEMBLES project*. Met Office Hadley Centre, Exeter. 160 pp.

von Storch H, Navarra A (eds). 1999. *Analysis of Climate Variability: Applications of Statistical Techniques*. Springer: Berlin.

Wang Y, Leung LR, McGregor JL, Lee DK, Wang WC, Ding Y, Kimura F. 2004. Regional Climate Modeling: Progress, Challenges, and Prospects. *Journal of the Meteorological Society of Japan* 82: 1599—1628.

Widmann M, Bretherton CS, Salathé E. 2003. Statistical Precipitation Downscaling over the Northwestern United States Using Numerically Simulated Precipitation as Predictor. *Journal of Climate* 16: 799—816.

Wilby R, Wigley T. 1997. Downscaling general circulation model output: a review of methods and limitations. *Progress in Physical Geography* 21: 530—548.

Wilks DS. 2006. *Statistical Methods in Atmospheric Science, Volume 91 of International Geophysics Series*. Academic Press: San Diego, London.

Wood AW, Leung LR, Sridhar V, Lettenmaier DP. 2004. Hydrologic Implications of Dynamical and Statistical Approaches to Downscale Climate Model Outputs. *Climatic Change* 62: 189—216.

INDEX

C

G

H

I

N

O

P

Q

R

S

T

U

V

W

X

Z